Photo Underwood & Underwood
CAPT. E. J. SMITH
The Commander of the Titanic, who went down with his ship

STEAMSHIP TITANIC

Wreck and Sinking
of the
TITANIC

==================
The Ocean's Greatest Disaster
==================

A Graphic and Thrilling Account of the Sinking of the greatest Floating Palace ever built, carrying down to watery graves more than 1500 souls.

Giving Exciting Escapes from death and acts of heroism not equalled in ancient or modern times, told by

THE SURVIVORS

Including History of Icebergs, the Terror of the Seas, Wireless Telegraphy and Modern Shipbuilding

==============================
ILLUSTRATED THROUGHOUT WITH PHOTOGRAPHS
AND DRAWINGS MADE EXPRESSLY FOR THIS BOOK
==========================

First published in 1912
First published in this edition in January 1998 by
Breese Books Ltd
164 Kensington Park Road, London W11 2ER, England
Reprinted February 1998

© Breese Books Ltd, 1998

All rights reserved

No parts of this publication may be reproduced, stored in retrieval systems or transmitted in any form or by any means, electronic, mechanical, photocopying, recording or otherwise, except brief extracts for the purposes of review, without prior permission of the publishers.

Any paperback edition of this book, whether published simultaneously with, or subsequent to the casebound edition, is sold subject to the condition that it shall not by way of trade, be lent, resold, hired out or otherwise disposed of without the publisher's consent, in any form of binding or cover other than that in which it was published.

ISBN: 0 947533 621

All illustrations reproduced by kind permission of
Retrograph Archive, London

Printed and bound in Great Britain by
Itchen Printers Ltd, Southampton

FACTS ABOUT THE *TITANIC*

The *Titanic's* length over all was 882 feet 6 inches. 182½ feet more than the height of the Metropolitan tower in New York City, and 3 1-3 times the height of Chicago's highest building. The Bunker Hill monument is one-fourth as high, and the Washington monument itself 300 feet shorter.

Some of the statistics follow:

Tonnage, registered	45,000
Tonnage, displacement	66,000
Length over all	882 feet, 6 inches
Breadth over all	92 feet, 6 inches
Breadth over boat deck	94 feet
Height from bottom of keel to boat deck	97 feet, 4 inches
Height from bottom of keel to top of captain's house	105 feet, 7 inches
Height of funnels above casing	72 feet
Height of funnels above boat deck	81 feet, 6 inches
Distance from top of funnel to keel	175 feet
Number of steel decks	11
Number of watertight bulkheads	15
Passengers carried	2,500
Crew	860
Cost	$10,000,000

Every line was calculated to be a little more impressive than that on any ship previously built. The great steel plates used in the hull included some as long as 36 feet, weighing 4½ tons each. Some of the great steel beams were 92 feet long, weighing 4 tons.

The rudder itself weighed 100 tons and of course was operated by electricity. The center turbine weighed 22 tons, and each of the two wing propellers 38 tons. The big boss arms from which the propellers were suspended tipped 73 tons. Even the anchor chains contributed their dimensions to the amazing total, with each link tipping 175 pounds. The 3,000,000 rivets used in construction weighed in aggregate 1,200 tons.

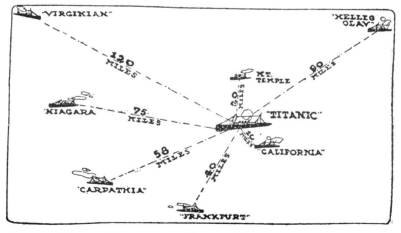

DIAGRAM SHOWING LOCATION AND DISTANCE FROM TITANIC OF OTHER SHIPS ON NIGHT OF DISASTER

LIST OF THE DEAD

The following list of passengers missing from the *Titanic*, revised from last reports from the *Carpathia*, contains only 914 actual names out of the total of 1,635 lost, but many more are accounted for in the steerage reports under the word "family." Still more of the victims in the steerage have not yet been named, and few, if any, of the names of the missing among the crew have been reported.

FIRST CABIN

Anderson, Harry.
Allison, H. J.
Allison, Mrs., and maid
Allison, Miss.
Andrews, Thomas.
Artagavoytia, Ramon.
Astor, Col. J. J., and servant.
Anderson, Walker.
Beattie, T.
Brandies, E.
Mrs. Wm. Bucknell's maid.
Baumann, J.
Baxter, Mr. and Mrs. Quigg.
Bjornstrom, H.
Birnbaum, Jacob.
Blackwell, S. W.
Borebank, J. J.
Bowden, Miss.
Brady, John B.
Brewe, Arthur J.
Butt, Major A.
Clark, Walter M.
Clifford, George Q.
Colley, E. P.
Cardeza, T. D. M., servant of.
Cardeza, Mrs. J. W., maid of.
Carlson, Frank.
Case, Howard B.
Cavendish, W. Tyrrell.
Corran, F. M.
Corran, J. P.
Chafee, Mr. H. I.
Chisholm, Robert.
Compton, A. T.
Crafton, John B.
Crosby, Edward G.
Cumings, J. Bradley.
Davidson, Thornton.
Dulles, William C.
Douglas, W. D.
Nurse of Douglas,
Master R.
Eustis, Miss E. M., may be reported saved as Miss Ellis.
Evans, Miss E.
Fortune, Mark.
Foreman, B. L.
Fortune, Charles.
Franklin, T. P.
Futrelle, J.
Gee, Arthur.
Goldenberg, E. L.
Goldschmidt, G. B.
Greenfield, G. B.
Giglio, Victor.
Guggenheim, Benj.
Servant of Harper, Henry S.
Hays, Charles M.
Maid of Hays, Mrs. Charles M.
Head, Christopher.
Hilliard, H. H.
Hopkins, W. F.
Hogenheim, Mrs. A.
Harris, Henry B.
Harp, Mr. and Mrs. Charles M.
Harp, Miss Margaret, and maid.
Hoyt, W. F.
Holverson, A. M.
Isham, Miss A. E.
Servant of J. Bruce Ismay.
Julian, H. F.
Jones, C. C.
Kent, Edward A.
Kenyon, Mr. and Mrs. F. R. (may be reported saved as Kenchen and Kenyman).
Kimball, Mr. and Mrs. E. N. (may be reported saved as Mr.
and Mrs. E. Kimberley).
Klober, Herman.
Lambert, Williams.
Lawrence, Arthur.
Long, Milton.
Longley, Miss G. F.
Lewy, E. G.
Lindsholm, J. (may be reported saved as Mrs. Sigrid Lindstrom).
Loring, J. H.
Lingrey, Edward.
Maguire, J. E.
McCaffry, T.
McCaffry, T., Jr.
McCarthy, T., Jr.
Marvin, D. W.
Middleton, J. C.
Millett, Frank D.
Minahan, Dr. and Mrs.
Marechal, Pierre.
Meyer, Edgar J.
Molson, H. M.
Moore, C., servant.
Newall, Miss T.
Nicholson, A. S.
Ovies, S.
Ostby, E. C.
Ornout, Alfred T.
Parr, M. H. W.
Pears, Mr. and Mrs. Thomas.
Penasco, Mr. Victor.
Partner, M. A.
Payne, V.
Pond, F., and maid.
Porter, Walter.
Reuchlin, J.
Maid of Robert, Mrs. E.
Roebling, W. A., 2d.
Rood, Hugh R.
Roes, J. Hugo.
Maid of Cnts. Rothes.
Rothschild, M.
Rowe, Arthur.
Ryerson, A.
Shutes, Miss E. W. (probably reported saved as Miss Shuter).
Maid of Mrs. G. Stone.
Straus, Mr. and Mrs. Isidor.
Silvey, William B.
Maid of Mrs. D. C. Spedden.
Spedden, Master D., and nurse.
Spencer, W. A.
Stead, W. T.
Stehli, Mr. and Mrs. Max Frolisher.
Sutton, Frederick.
Smart, John M.
Smith, Clinch.
Smith, R. W.
Stewart, A. A. (may be reported saved as Frederick Stewart).
Smith, L. P.
Taussig, Mrs. Emil.
Maid of Mrs. Thayer.
Thayer, John B.
Thorne, C.
Vanderhoof, Wyckoff.
Walker, W. A.
Warren, F. M.
White, Percival A.
White, Richard F.
Widener, G. D., and servant.
Widener, Harry.
Wood, Mr. and Mrs. Frank P.
Weir, J.
Wick, George D.
Williams, Duane.
Wright, George.

LIST OF THE DEAD

SECOND CABIN

Abelson, Samson.
Andrew, Frank.
Ashby, John.
Aldworth, C.
Andrew, Edgar.
Beacken, James H.
Brown, Mrs.
Banfield, Fred.
Beight, Nail.
Braily, Bandsman.
Breicoux, Bandsman.
Bailey, Percy.
Bainbridge, C. R.
Byles, the Rev. Thos.
Beauchamp, H. J.
Bees, F. Lawrence.
Berg, Miss E.
Benthan, I.
Bateman, Robert J.
Butler, Reginald.
Botsford, Hall.
Bowcener, Solomon.
Berriman, William.
Clarke, Charles.
Clark, Bandsman.
Corey, Mrs.
Carter, Rev. Ernest.
Carter, Mrs.
Coleridge, Reginald.
Chapman, Charles.
Cunningham, Alfred.
Campbell, William.
Collyer, Harvey.
Corbett, Mrs. Irene.
Chrman, John R.
Chapman, Mrs. E.
Colander, Erie.
Cotterill, Harry.
Charles, Wm. (probably reported saved as Wm. Charles).
Deacon, Percy.
Davis, Charles (may be reported saved as John Davies).
Debben, William.
De Brits, Jose.
Danborny, H.
Drew, James.
Drew, Master M.
David, Master J. W.
Duran, Miss A.
Dounton, W. J.

Del Vario, S.
Del Vario, Mrs.
Enander, Ingvar.
Eitmiller, G. F.
Frost, A.
Fynnery, Mr.
Faunthrope, H.
Fillbrook, C.
Funk, Annie.
Fahlstrom, A.
Fox, Stanley N.
Greenberg, S.
Giles, Ralph.
Gaskell, Alfred.
Gillespi, William.
Gibert, William.
Gall, Harry.
Gall, S.
Gill, John.
Giles, Edgar.
Giles, Fred.
Goh, Harry.
Gale, Phadruch.
Garvey, Lawrence.
Hickman, Leonard.
Hickman, Lewis.
Hume, bandsman.
Hickman, Stanley.
Hood, Ambrose.
Hodges, Henry P.
Hart, Benjamin.
Harris, Walter.
Harper, John.

Harbeck, W. H.
Hoffman, Mr.
Hoffman, Child.
Hoffman, Child.
Herman, Mrs. S.
Howard, P.
Howard, Mrs. E. T.
Hale, Reginald.
Hamatainen, A., and infant son (probably reported saved as Anna Harnlin).
Hilunen, M.
Hunt, George.
Jacobson, Mr.
Jacobson, Mrs.
Jacobson, Sydney.
Jeffery, Clifford.
Jeffery, Ernest.

Jenkin, Stephen.
Jarvis, John D.
Keane, Daniel.
Kirkland, Rev. C.
Karnes, Mrs. F. G.
Keynaldo, Miss.
Krillner, J. H.
Krins, bandsman.
Knight, R.
Karines, Mrs.
Kantar, Selna.
Kantar, Mrs. (probably reported saved as Miriam Kanton.)
Lengam, John.
Levy, P. J.
Lahtigan, William.
Lauch, Charles.
Leyson, R. W. N.
Laroche, Joseph.
Lamb, J. J.
McKane, Peter.
Milling, Jacob.
Mantville, Joseph.
Malachard Noll, (may be reported saved as Mme. Mellcard).
Morawek, Dr.
Manglovacell, E.
McCrae, Arthur G.
McCrie, James M.
McKane, Peter D.
Mudd, Thomas.
Mack, Mary.
Marshall, Henry.
Mayberg, Frank H.
Meyer, August.
Myles, Thomas.
Mitchell, Henry.
Matthews, W. J.
Nessen, Israel.
Nicholls, Joseph C.
Norman, Robert D.
Nasser, Nicholas (may be reported saved as Mrs. Nasser).
Otteo, Richard.
Phillips, Robert.
Ponesell, Martin (may be reported saved as M. F. Pososons).
Pain, Dr. Alfred.
Parkes, Frank.

Pengelly, F.
Pernot, Rene.
Peruschitz, the Rev.
Parker, Clifford.
Paulbaum, Frank.
Rogers, Getina (probably reported saved as Miss E. Rogers).
Renouf, Peter E.
Rogers, Harry.
Reeves, David.
Slemen, R. J.
Sjoberg, Hayden.
Slatter, Miss H. M.
Stanton, Ward.
Sinkkonen, A. (probably reported saved as Anna Sinkkanea).
Sword, Hans K.
Stokes, Philip J.
Sharp, Percival.
Sedgwick, Mr.
Smith, Augustus.
Sweet, George.
Sjostedt, Ernst.
Tocmey, Ellen (may be reported saved as Ellen Formery).
Taylor, bandsman.
Turpin, William.
Turpin, Mrs. Dorothy.
Turner, John H.
Trouneansky, M.
Tervan, Mrs. A.
Trant, Mrs. Jesse (probably reported saved as Mrs. Jessie Traut).
Veale, James.
Wilhelm, Chas. (probably reported saved as Chas. Williams.)
Watson, E.
Woodward, bandsman.
Ware, William C.
Weiz, Leopold.
Wheadon, Edward.
Ware, John J.
Ware, Mrs. (may be reported saved as Miss F. Mare.)
West, E. Arthur.
Wheeler, Edwin.
Wenman, Samuel.

THIRD CLASS—STEERAGE

Allum, Owen.
Alexander, William.
Adams, J.
Alfred, Evan.
Allen, William.
Akar, Nourealain.
Assad, Said.
Alice, Agnes.
Abbing, Anthony.
Aks, Tilly.
Attala, Malakka.
Ayont, Bancura.
Ahmed, Ali.
Alhomaki, Ilmari.
Ali, William.
Anders, Gustafson.
Assam, Ali.

Asin, Adola.
Anderson, Albert.
Anderson, Ida.
Anderson, Thor.
Aronson, Ernest.
Ahlin, Johanna.
Anderson, Anders, and family.
Anderson, Carl.
Anderson, Samuel.
Andressen, Paul.
Augustan, Albert.
Abelsett, Olaf.
Adelseth, Karen.
Adolf, Humblin.
Anderson, Erna.
Angheloff, Minko.

Arnold, Josef.
Arnold, Josephine.
Asplund, Johan.
Braun, Lewis.
Braun, Owen.
Bowen, David.
Beavan, W.
Bachini, Zabour.
Belmentoy, Hassef.
Badt, Mohamet.
Betros, Yazbeck.
Barry, ———.
Buckley, Katharine.
Burke, Jeremiah.
Barton, David.
Brocklebank, William.
Bostandyeff, Cuentche.

Benson, John.
Billiard, A., and two children.
Bontos, Hanna.
Baccos, Boulos.
Bexrous, Tannous.
Burke, John.
Burke, Catharine.
Burke, Mary.
Burns, Mary.
Berglind, Ivar.
Balkie, Cerin.
Rrobek, Carl.
Backstrom, Karl.
Berglund, Hans.
Bjorkland, Ernest.
Can, Ernest.

LIST OF THE DEAD

THIRD CLASS—STEERAGE (CONTINUED)

Crease, Ernest.
Cohett, Gershon.
Coutts, Winnie, and two children.
Cribb, John.
Cribb, Alice C.
Catavelas, Vassilios.
Caram, Catharine.
Cannavan, P.
Carr, Jenny.
Chartens, David.
Conline, Thomas.
Celloti, Francesco.
Christmann, Emil.
Coxon, Daniel.
Corn, Harry.
Carver, A.
Cook, Jacob.
Chip, Chang.
Chaaini, Georges.
Chronopolous, D.
Connaghton, M.
Conners, P.
Carls, Anderson.
Carlsson, August.
Coelhe, Domingo.
Carlson, Carl.
Coleff, Sotie.
Coleff, Peye.
Cor, Ivan, and family.
Calic, Manda.
Calic, Peter.
Cheskosic, Luka.
Cacic, Gego.
Cacic, Luka.
Cacic, Taria.
Carlson, Julius.
Crescovic, Maria.
Dugemin, Joseph.
Dean, Bertram.
Dorkings, Edward.
Dennis, Samuel.
Dennis, William.
Drazenovic, Josef.
Daher, Shedid.
Daly, Eugene.
Dwar, Frank.
Davies, John.
Dowdell, E.
Davison, Thomas.
Davison, Mary.
Dahl, Charles.
Drapkin, Jennie.
Donahue, Bert.
Doyle, Ellen.
Dwyer, Tillie.
Dakic, Branko.
Danoff, Yoto.
Dantchoff, Christo.
Denkoff, Mitto.
Dintcheff, Valtcho.
Dedalic, Regzo.
Dahlberg, Gerda.
Demossemacker, E.
Demossemacker, G.
Dimic, Jovan.
Dahl, Mauritz.
Dalbom, E., and fam.
Dyker, Adolph.
Dyker, Elizabeth.
Everett, Thomas.
Empuel, Ethel.
Elsbury, James.
Elias, Joseph.
Elias, Joseph.
Enas, Hannah.

Elias, Foofa.
Emmet, Thomas.
Ecimozic, Joso.
Edwarlson, Gustave.
Eklund, Hans.
Ekstrom, Johan.
Ford, Arthur.
Ford, M., and family.
Franklin, Charles.
Foo, Cheong.
Farrell, James.
Flynn, James.
Flynn, John.
Foley, Joseph.
Foley, William.
Finote, Lingi.
Ficcher, Eberhard.
Goodwin, F., and fam.
Goldsmith, F., and family.
Guest, Frank.
Green, George.
Ca..rth, John.
Conski, Leslie.
Ghorgeff, Stano.
Ghemat, Emar.
Gerios, Youssef.
Gerios, Assaf.
Ghalil, Saal.
Gallagher, Martin.
Ganavan, Mary.
Glinagh, Katie.
Glynn, Mary.
Gronnestad, Daniel.
Gustafsch, Gideon.
Goldsmith, Nathan.
Goncalves, Mancel.
Gustafson, Johan.
Graf, Elin.
Gustafson, Alfred.
Hyman, Abraham.
Harknett, Alice.
Hane, Youssef, and two children.
Haggendon, Kate.
Haggerty, Nora.
Hart, Henry.
Howard, May.
Harmer, Abraham.
Hachini, Najib.
Helene, Eugene.
Healy, Nora.
Henery, Della.
Hemming, Nora.
Hansen, Claus.
Hansen, Fanny.
Heininan, Wendla.
Hervonen, Helga, and child.
Haas, Alaisa.
Hakkurainen, Elin.
Hakkurainen, Pekka.
Hankomen, Eluna.
Hansen, Henry.
Hendekovic, Ignaz.
Hickkinen, Laina.
Holm, John.
Hadman, Oscar.
Eaglund, Conrad.
Haglund, Ingvald.
Henriksson, Jenny.
Hillstrom, Hilda.
Holten, Johan.
Ing, Hen.
Iemenen, Manta.
Ilmakangas, Pista.

Ilmakangas, Ida.
Ilieff, Kriste.
Ilieff, Ylio.
Ivanhoff, Kanie.
Johnson, A., and fam.
Jamila, N., and child.
Jenymin, Annie.
Johnstone, W.
Joseph, Mary.
Jeannasr, Hanna.
Johannessen, Berdt.
Johannessen, Elias.
Johansen, Nils.
Johanson, Oscar.
Johansson, Gustav.
Johkoff, Lazer.
Johnson, E., and fam.
Johnson, Jakob.
Johnsson, Nils.
Jansen, Carl.
Jardin, Jose.
Jensen, Hans.
Johansson, Eric.
Jussila, Eric.
Jutel, Henry.
Johnsson, Carl.
Jusila, Katrina.
Jusila, Maria.
Keefe, Arthur.
Kassen, Houssent.
Karum, F., and child.
Kelly, Anna.
Kelly, James.
Kennedy, John.
Kerane, Andy.
Kelley, James.
Keeni, Fahim.
Khalil, Lahia.
Kiernan, Philip.
Kiernan, John.
Kilgannon, Theo.
Kakic, Tido.
Karajis, Milan.
Karkson, Einar.
Kalvig, Johannes.
King, Vin., and fam.
Kallio, Nikolai.
Karlson, Nils.
Klasson, K., two chil.
Lovell, John.
Lob, William.
Lobb, Cordelia.
Lester, James.
Lithmau, Simon.
Leonard, I.
Lemberopolous, P.
Lakarian, Orsen.
Lane, Patrick.
Lennon, Dennis.
Lam, Ah.
Lam, Len.
Lang, Fang.
Ling, Lee.
Lockyer, Edward.
Latife, Maria.
Lennon, Mary.
Linehan, Michael.
Leinenen, Antti.
Lindell, Edward.
Lindell, Elin.
Lindqvist, Vine.
Larson, Viktor.
Lefebre, F., and fam.
Lindblom, August.
Lulic, Nicola.
Lundal, Hans.

Lundstrom, Jan.
Lyntakoff, Stanke.
Landegren, Aurora.
Laitinen, Sotia.
Larsson, Bengt.
Lasson, Edward.
Lindahl, Anna.
Lundin, Olga.
Moore, Leonard.
Mackay, George.
Meek, Annie.
Mikalsen, Sander.
Miles, Frank.
Miles, Frederick.
Morley, William.
McNamee, Neal.
McNamee, Ellen.
Meanwell, Marian.
Meo, Alfonso.
Maizner, Simon.
Murdlin, Joseph.
Moore, Belle.
Moor, Meier.
Maria, Joseph.
Mantour, Mousea.
Moncarek, O., 2 chil.
McElroy, Michael.
McGowan, Katharine.
McMahon, ———.
McMahon, Martin.
Madigan, Maggie.
Manion, Margaret.
Mechan, John.
Mocklare, Ellis.
Moran, James.
Mulvihill, Bertha.
Murphy, Kate.
Mikanen, John.
Melkebuk, Philemon.
Merms, Leon.
Midtsjo, Carl.
Myhrman, Oliver.
Myster, Anna.
Makinen, Kale.
Mustafa, Nasr.
Mike, Anna.
Mustmans, Fatina.
Martin, Johan.
Malinoff, Nicola.
McCoy, Bridget.
Markoff, Martin.
Marinko, Dimitri.
Mineff, Ivan.
Minkoff, Lazar.
Mirko, Dika.
Mitkoff, Nitto.
Moen, Sigurd.
Nancarror, William.
Nomagh, Robert.
Nakle, Trotik.
Naked, Maria.
Nosworthy, Richard.
Naughton, Hannah.
Norel, Manseur.
Niels, ———.
Nillson, Herta.
Nyoven, Johan.
Naidenoff, Penke.
Nankoff, Minko.
Nedelic, Petroff.
Nenkoff, Christie.
Nilson, August.
Nirva, Isak.
Nandewallo, Nestor.
O'Brien, Dennis.
O'Brien, Hanna.

LIST OF THE DEAD

THIRD CLASS—STEERAGE (CONTINUED)

O'Brien, Thomas.
O'Donnell, Patrick.
Odele, Catharine.
O'Connor, Patrick.
O'Neill, Bridget.
Olsen, Carl.
Olsen, Ole.
Olson, Elin.
Olson, John.
Ortin, Amin.
Odahl, Martin.
Olman, Velin.
Olsen, Henry.
Olman, Mara.
Olsen, Elide.
Orescovic, Teko.
Pedruzzi, Joseph.
Perkin, John.
Pearce, Ernest.
Peacock, T., two chil.
Potchett, George.
Peterson, Marius.
Peters, Katie.
Paulsson, A., and fam.
Panula, M., and fam.
Pekonami, E.
Peltomaki, Miheldi.
Pacruic, Mate.
Pacruic, Tamo.
Pastche, Petroff.
Pietcharsky, Vasil.
Palovic, Vtefo.
Petranec, Matilda.
Person, Ernest.
Pasic, Jacob.
Planke, Jules.
Peterson, Ellen.
Peterson, Olaf.
Peterson, Wohn.
Rouse, Richard.
Rush, Alfred.

Rogers, William.
Reynolds, Harold.
Riordan, Hannah.
Ryan, Edward.
Rainch, Razi.
Roufoul, Aposetun.
Read, J mes.
Robins, Alexander.
Robins, Charity.
Risian, Samuel.
Risian, Emma.
Runnestvet, Kristian.
Randeff, Alexandre.
Riutamaki, Matti.
Rosblom, H., and fam.
Ridegain, Charles.
Sadowitz, Harry.
Saundercock, W.
Sage, Jno., and fam.
Sawyer, Frederick.
Spinner, Henry.
Shorney, Charles.
Sarkis, Lahound.
Sultani, Meme.
Stankovic, Javan.
Salini, Antoni.
Seman, Betros.
Sadlier, Matt.
Scanlon, James.
Shaughnessay, P.
Simmons, John.
Serota, Maurice.
Somerton, F.
Slocovski, Selmen.
Sutchall, Henry.
Sather, Simon.
Storey, T.
Spector, Woolf.
Sirayman, Peter.
Samaan, Jouseef.

Saiide, Barbara.
Saad, Divo.
Sarkis, Madiresian.
Shine, Ellen.
Sullivan, Bridget.
Salander, Carl.
Sepelelanaker, Alfons.
Skog, Wm., and fam.
Solvang, Lena.
Stranberg, Ida.
Strilik, Ivan.
Salonen, Ferner.
Sivic, Husen.
Svenson, Ola.
Svedst, ———.
Sandman, Mohan.
S::ljilsvick, Anna.
Schelp, Peter.
Sl vola, Antti.
Slabenoff, Peter.
Staneff, Ivan.
Stoytcho, Mikoff.
Stoyte off, Ilia.
Sydcoff, Todor.
Sandstrom, Agnes, and two children.
Sheerlinch, Joan.
Smiljanik, Mile.
Strom, E., and child.
Svensson, John.
Swensson, Edwin.
Tobin, Roger.
Thomson, Alex.
Theobald, Thomas.
Tomlin, Ernest.
Thorncycroft, P.
Thorneycroft, F.
Torber, Ernest.
Trembisky, Berk.
Tilley, Edward.
Tamini, Hilion.

Tannans, Daper.
Thomas, John.
Thomas, Charles.
Thomas, Tannous.
Tumin, T., and infant.
Tikkanen, Juho.
Tonglin, Gunner.
Turoin, Stefan.
Turgo, Anna.
Tedoreff, Ialie.
Usher, Haulmer.
Nzelas, Jose.
Vander and family.
Vereruysse, Victor.
Vjoblom, Anna.
Vaciens, Adulle.
Vandersteen, Leo.
Vanimps, J., and fam.
Vatdevehde, Josep.
Williams, Harry.
Williams, Leslie.
Ware, Frederick.
Warren, Charles.
Waika, Said.
Wazli, Jousef.
Wiseman, Philip.
Werber, James.
Windelor, Einar.
Weller, Edward.

Wendal, Olaf.
Wistrom, Hans.
Wiklund, Jacob.
Wiklund, Carl.
Wenzel, Zinhart.
Wirz, Albert.
Wittewrongel, Camille.
Youssef, Brahim.
Yalsevac, Ivan.
Zakarian, Mapri.
Zievens, Rene.
Zimmerman, Leo.

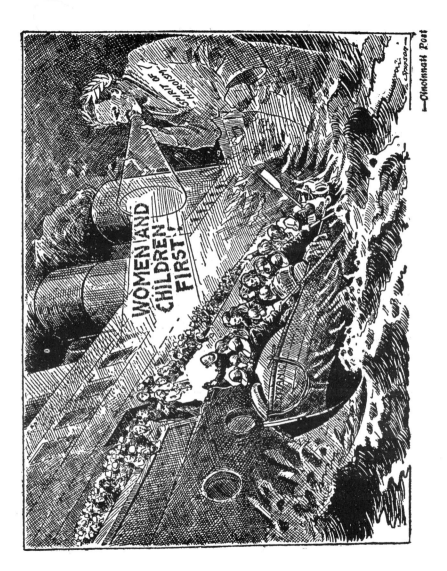

Copyright by Campbell Studio, N. Y.
COLONEL JOHN JACOB ASTOR
Lost with the Titanic, and his young bride, who was rescued

Story of the Wreck of the Titanic

CHAPTER I

THE TWO TITANS

AS the *Titanic* drew away from the wharf to begin her only voyage, a common emotion quickened the thousands who were aboard her. Grimy slaves who worked and withered deep down in the glaring heat of her boiler rooms, on her breezy decks men of achievement and fame and millionaire pleasure seekers for whom the boat provided countless luxuries, in the steerage hordes of emigrants huddled in straited quarters but with their hearts fired for the new free land of hope; these, and also he whose anxious office placed him high above all—charged with the keeping of all of their lives—this care-furrowed captain on the bridge, his many-varied passengers, and even the remotest menial of his crew, experienced alike a glow of triumph as they faced the unknown dangers of the deep, a triumph born of pride in the enormous, wonderful new ship that carried them.

For she was the biggest boat that ever had been in the world. She implied the utmost stretch of construction, the furthest achievement in efficiency, the bewildering embodiment of an immense multitude of luxuries for which only the richest of the earth could pay. The cost of the *Titanic* was tremendous—it had taken many millions of dollars—many months to complete her. Besides (and best of all) she was practically unsinkable her owners said; pierce her hull anywhere, and behind was a watertight bulkhead, a sure defense to flout the floods and hold the angry ocean from its prey.

Angry is the word—for in all her triumph of perfection the *Titanic* was but man's latest insolence to the sea. Every article in her was a sheer defiance to the Deep's might and majesty. The ship is not the ocean's bride; steel hull and mast, whirling shaft and throbbing engine-heart (products, all, of serviceable wonderworking fire)—what kinship have these with the wild and watery waste? They are an affront and not an affinity for the cold and alien and elusive element that at all times threatens to overwhelm them.

But no one on the *Titanic* dreamed of danger when her prow was first set westward and her blades began the rhythmic beat that must not cease until the Atlantic had been crossed. Of all the statesmen, journalists, authors, famous financiers who were among her passengers (many of whom had arranged their affairs especially to secure passage in this splendid vessel), in all that brilliant company it may be doubted if a single mind secreted the faintest lurking premonition of a fear. Other ships could come safely and safely go, much more this monster—why, if an accident occurred and worse came to worst, she was literally too *big* to sink! Such was the instinctive reasoning of her passengers and crew, and such the unconsidered opinion of the world that read of her departure on the fatal day which marked the beginning of her first voyage and her last.

No doubt her very name tempted this opinion: *Titanic* was she titled—as though she were allied to the famous fabled giants of old called Titans, who waged a furious war with the very forces of creation.

Out she bore, this giant of the ships, then, blithely to meet and buffet back the surge, the shock, of ocean's elemental might; latest enginery devised in man's eternal warfare against nature, product of a thousand minds, bearer of myriad hopes. And to that unconsidered opinion of the world she doubtless seemed even arrogant in her plenitude of power, like the elements she clove and rode—the sweeping winds above, the surging tide below. But this would be only in daytime, when the *Titanic* was beheld near land, whereon are multitudes of things beside which this biggest of the ships loomed large. When we imagine her alone, eclipsed by the solitude and immensity of night, a gleaming speck—no more—upon the gulf and middle of the vasty deep, while her gayer guests are dancing and the rest are moved to mirth or wrapped in slumber or lulled in security—when we think of her thus in her true relation, she seems not arrogant of power at all; only a slim and alien shape too feeble for her freight of precious souls, plowing a tiny track across the void, set about with silent forces of destruction compared to which she is as fragile as a cockle shell.

Against her had been set in motion a mass for a long time mounting, a century's stored-up aggregation of force, greater than any man-made thing as is infinity to one. It had expanded in the patience of great solitudes. On a Greenland summit, ages ago, avalanches of ice and snow collided, welded and then moved, inches in a year, an evolution that had naught to do with time. It was the true inevitable, gouging out a valley for its course, shouldering the precipices from its path. Finally the glacier reached the open Arctic, when a mile-in-width of it broke off and floated swinging free at last.

Does Providence directly govern everything that is? And did the Power who preordained the utmost second of each planet's journey, rouse up the mountain from its sleep of snow and send it down to drift, deliberately direct, into the exact moment in the sea of time, into the exact station in the sea of waters, where danced a gleaming speck—the tiny *Titanic*—to be touched and overborne?

It is easy thus to ascribe to the Infinite the direction of the spectacular phenomena of nature; our laws denote them "acts of God"; our instincts (after centuries of civilization) still see in the earthquake an especial instance of His power, and in the flood the evidence of His wrath. The floating menace of the sea and ice is in a class with these. The terror-stricken who from their ship beheld the overwhelming monster say that it was beyond all imagination vast and awful, hundreds of feet high, leagues in extent, black as it moved beneath no moon, appallingly suggestive of man's futility amidst the immensity of creation. See how, by a mere touch— scarcely a jar—one of humanity's proudest handiworks, the greatest vessel of all time, is cut down in her course, ripped up, dismantled and engulfed. The true Titan has overturned the toy.

Oh, where is now the boasted strength of that great hull of steel! Pitted against the iceberg's adamant it crumples and collapses. What of the ship unsinkable; assured so by a perfected new device? settling in the sea, shuddering to an inrush and an outburst of frigid water and exploding steam! All the effort of the thousand busy brains that built her, all the myriad hopes she bore—down, quite down! A long farewell to the toy Titan as the erasing waters fill and flatten smooth again to ocean's cold obliterating calm the handsbreadth she once fretted and defied!

Yes, it is easy to see God only in the grander manifestations of nature; but occasionally we are stricken by his speaking in the still small voice. Hundreds on this night of wreck were thus impressed. As the great steel-strong leviathan sank into the sea, those in the fleeing lifeboats heard, amid the thunder and the discord of the monster's breaking-up, afar across the waters floating clear, a tremulous insistence of sweet sound, a hymn of faith—utterly triumphant o'er the solitudes! Men had left their work to perish and turned themselves to God.

When he builds and boasts of his *Titanics,* man may be great, but it is only when he is stripped of every cloying attribute of the world's pomp and power that he can touch sublimity. Those on the wreck had mounted to it from the time the awful impact came. The rise began when men of intellect and noted works, of titled place and honored station, worked as true yoke-fellows with the steerage passengers to see that all the women and their little ones were safely placed within the boats. They did this calmly, while the steamer settled low and every instant brought the waters nearer to their breath; exulting as each o'erburdened lifeboat safely drew away, and cheering until the iceberg echoed back the sound. There was very little fear displayed; calm intrepidity was here the mark of a high calling. Captain Smith, indeed, was afraid, but it was only for the precious beings under God committed to his care. And how

WRECK OF THE TITANIC

manfully he minimized at first the danger until the rising surges creeping o'er the decks betrayed the awful truth. Then was the panic time! what cries were heard! what partings had and fond farewells! What love was lavished in renunciation and in life-and-death constancy when husband and wife refused to be separated in the hour that meant the inevitable death of one. But through all the time of terror the heroes of the *Titanic* remained true, nor yielded hearts to fear; and then, when all was done, when the last well-laden boat had safely put away, when the chill waters could be felt encroaching in the darkness, those who voluntarily awaited death, who had exemplified the sacred words: "Greater love hath no man than this, that a man lay down his life for a friend"—then these put heroism behind them for humility, rose to the greater height, threw themselves on Him who walked the waters to a sinking ship, as they sang in ecstasy the simple hymn of steadfast faith: "Nearer, my God, to Thee, nearer to Thee"!

Thus did man assert once more his high superiority among created things—he alone has power to revert to the unseen author of them all. Though compassed about with vast unfriendly Titans of the elements, builder himself of Toy Titans, like the boasted ship, that exist at the mercy of the sea and sky—at every fresh disaster that brings to nothingness his chiefest works, his spirit yet allies itself peculiarly with the power that only may be imagined and not seen; being persuaded that neither death, nor life, nor angels, nor principalities, nor powers, nor things present, nor things to come, nor height, nor depth, nor any other creature, shall be able to separate him from the love of God.

FRED S. MILLER.

—*Cleveland Plain Dealer*

CHAPTER II

STORY OF THE TITANIC

THE "UNSINKABLE" TITANIC STRIKES AN ICEBERG AND SINKS—HUNDREDS CARRIED TO SUDDEN AND UNTIMELY DEATH FOR LACK OF ADEQUATE LIFE-SAVING SERVICE—THE FACTS OF THE WRECK.

The mighty ship *Titanic,* the triumph of the shipbuilders, thronged with happy, confident people, interested in her first voyage and her speed record, ploughed her swift way across the Atlantic, which lay smooth and calm and clear. In the midst of pleasant amusements and happy dreams there came a slight shock, a glancing blow from an iceberg, a few minutes of calm disbelief—then horror incredible. The Titan of nature and the *Titanic* of mechanical construction had met in mid-ocean. The iceberg ripped open the ship's side, exposing her boilers to the icy water, causing their explosion, plunging hundreds of people to their death within the short space of two hours. This is the tragic story of the beautiful ocean palace that sailed forth so gallantly from harbor on her maiden trip, April 10, 1912,—buried under 2,000 fathoms of water with some 1,595 of her ill-fated passengers.

No more thrilling or pitiful tale has ever been written on the page of history—no greater record of human sacrifice and heroism.

The Titanic was the last word in ship building and she set forth on her first voyage, the pride of an admiring world. Her luxurious appointments were beyond criticism, beautiful salons, reading and lounging rooms, palm courts, Turkish baths, private baths, a gymnasium, a swimming pool, a ballroom and billiard hall, everything one could imagine as making for comfort. Her mechanical construction was thought to be as perfect, and in the minds of her passengers was a faith in her "unsinkable" character almost unshakable. She carried nearly a full passenger roll, 2,340 people including the crew, as generally estimated, and was provided with only twenty lifeboats, sixteen ordinary lifeboats and four collapsible boats—accommodation for about a third of her passengers. These numbered some of the wealthiest and most prominent people on both sides of the Atlantic, John Jacob Astor, Major Archibald Butt, Benjamin Guggenheim, Isidor Straus, Charles M. Hays, Arthur Ryerson, Henry B. Harris, William T. Stead, Jacques Futrelle, and many more who gave up their lives in common with the humblest passenger in the steerage.

After the usual concert, Sunday evening, April 14, the passengers were in the midst of retiring or were

WRECK OF THE TITANIC

amusing themselves in card and reading rooms. Some few were on deck enjoying the splendid evening, clear and fair, the ocean wonderfully calm. Suddenly there came a slight rocking of the ship, so slight as to be unnoticed by many. "Grazed an iceberg. Nothing serious," was the general comment as men resumed their interrupted card games. That was 11:40 P. M. Many people went to bed without another thought. The berg had been sighted only a quarter of a mile away, too late to check the ship's speed, so she rushed into the mass of ice, projecting only about eighty feet above sea level but reaching dangerously into the depths. The shock of the blow was so slight as to be scarcely perceptible to the unconscious passengers. But nevertheless it was a stroke dealing out death. For the *Titanic,* pushed on by her tremendous momentum of 21 knots an hour, sliding against the knife-like ledge, projecting unseen into the water, ripped her side open on the ice, shattering her air-tight bulkheads. This permitted her gradual sinking, thereby allowing the icy waters to penetrate to her boilers, which had been working at high pressure, and causing their explosion, sending her to the bottom within two and one-half hours from the time she struck the iceberg.

Captain Smith took command as soon as the ship struck and the engines were stopped instantly. This sudden cessation of the constant vibration drew the passengers' attention more than did the shock of the collision. Life belts were ordered on the people immediately, and the boats were made ready, though the passengers thought all the time it was merely done for the sake of extraordinary precaution.

In the first boat the occupants were nearly all men, for there were no women on deck. The stewards and stewardesses were ordered below to summon the people from their staterooms, and when they came rushing out, some in their night clothes, some in evening gowns, all startled at the order but even yet believing in the strength of the *Titanic,* the rule "women first" was rigidly enforced. Unwillingly the women were torn from their husbands, or placed in the boats by their husbands with the assurance that they would follow in other boats. In this way the boats were loaded with women and children, protesting but passive for the most part, with just two or three men to manage the oars. The scene was one of remarkable order. There was no mad struggle for safety; the men stood back and sent the women out, with very little disturbance. The report was circulated that the men and women were to be put in separate boats; also that there were boats on the other side of the ship and they were simply going later. Many thought, too, that their boats would soon be called back—that it was a mere matter of a short side-excursion. So the boats were lowered away, and only when they were out in the water did their occupants realize the real danger. Then they could see the desperate plight of the *Titanic*.

As the *Titanic* sank gradually the water reached her engines, and an explosion tilted her decks, the list becoming more pronounced and consequently more dangerous every moment. Still the boats were loaded with women and children, until the last one swung off just in time.

The doomed multitude remaining shared her fate. Some leaped into the sea and clutched at floating wreckage; some sank with her, swimming to bits of wreckage as they struck the water; most of these were drowned, though a few escaped miraculously, picked up by the lifeboats or keeping themselves afloat by means of drifting boards and ship furnishings.

As the ship went down at 2:20 Monday morning, her colors flying, her captain in his place on the bridge, her bulk aglow with twinkling lights, the majority of her passengers looking out to sea from her decks, her string band playing "Nearer, My God, to Thee," united for the final moment the souls of the unhappy ones in safety of the frail boats with those loved ones helplessly going to their death.

Then the lights winked, the black mass surged under and the death cries of the hundreds broke into the quiet night.

That was soon over, but the suffering in the lifeboats continued for hours. It was bitterly cold, due to the proximity of the iceberg; many of the boats were dashed partly full of the icy water; none of their occupants were sufficiently clad. In some of the boats, the women had to take the oars and they rowed with bleeding hands, these delicately nurtured ladies who proved their claim on heroism equal to that of the gentlemen. The boats were not provided with food, water, lighting facilities, necessities of any kind, and when the *Carpathia,* summoned by wireless, reached them, they could only signal by means of fragmentary letters and matches found about the persons of some of the passengers.

For four long hours they floated about, dazed by sorrow, nearly insensible from the bitter exposure to cold and wet, until the good ship *Carpathia* picked them up. Once in her cabins, they were given food and clothes; warmed, but not comforted. After the rescue, a service of thanksgiving, funeral service for the lost, was held—one of the most heart-breaking scenes ever enacted.

Thus ended the career of the *Titanic,* but her story will live long in the hearts of the bereft survivors, and, to all the world, it bears a message that cannot be ignored—the message that to the god of commercial greed human sacrifices shall not be allowed at sea.

When the gallant ship *Titanic,* fair and false, set

forth on her initial trip with her 2,340 passengers, they little dreamed they were destined to point a moral to the world—that they were to be the instruments to demonstrate the criminal negligence of ship builders in deliberately sending forth vessels luxuriously equipped with every convenience and comfort, except the most essential one—lifeboats.

This great ocean liner—representing the acme of ship construction—went to her ruin after striking a huge iceberg in her course, an accident which probably was unavoidable, though greater care might have been exercised in the matter of speed.

To the twenty frail lifeboats fell the burden of keeping her 2,340 passengers afloat until the inevitable help should come, with the equally inevitable result that only 745 people emerged from the ill-fated wreck.

The cause for the disaster is undeniable; the reason for the loss of life is equally clear. The tales of horror of the survivors point to one single ominous fact; lack of adequate, commonsense protection of life paid to the Atlantic sea bottom the horrid toll of 1,595 persons.

Unequalled in their terrible, thrilling quality, the stories of this disaster; the striking of the iceberg, the loading of the boats, the agonized farewell, the mad leaps into the sea, the fearful hours upon the water before rescue, and the bitter revelations of those lost, all these things stir the heart to sympathy and the conscience to a demand for lawful, law enforced safeguards that shall prevent another such grim tragedy.

These murdered hundreds were merely another instance of the innocent sacrifices offered to the god of commercial profit. Some day, it is written, we shall cease this heathen worship; we shall demand proper precautions for our people, even though it be at the expense of a few paltry dollars. The time is now.

Laws shall be made and laws shall be enforced, and the future millions shall go to sea in ships provided with adequate safeguards. This is the service performed for us by these martyrs of the *Titanic*.

—*Cleveland Plain Dealer*

WAITING IN SUSPENSE

CHAPTER III

SPUR OF ICEBERG RIPPED OPEN BOTTOM OF THE *TITANIC*

GIGANTIC VESSEL LITERALLY DISEMBOWELED BY SUBMERGED FLOE WHILE SPEEDING—LITTLE SHOCK WAS FELT—PASSENGERS FOR HALF AN HOUR BELIEVED DAMAGE WAS SLIGHT AND TOOK THINGS CALMLY—MANY WERE IN THEIR STATEROOMS.

It was the submerged spur of an iceberg of ordinary proportions that sent the White Star liner *Titanic* more than two miles to the bottom of the Atlantic off the banks of Newfoundland. The vessel was steaming almost full tilt through a gently swelling sea and under a starlit sky, in charge of First Officer Murdock, who a moment after the collision surrendered the command to Capt. Smith, who went down with his boat.

The lifeboats that were launched were not filled to their capacity. The general feeling aboard the ship was, even after the boats had left its sides, that the vessel would survive its wound, and the passengers who were left aboard believed almost up to the last moment that they had a chance for their lives.

The captain and officers behaved with the utmost gallantry, and there was perfect order and discipline in the launching of the boats, even after all hope had been abandoned for the salvation of the ship and of those who were on board.

PLACID SEA HID DEATH

The great liner was plunging through a comparatively placid sea on the surface of which there was much mushy ice and here and there a number of comparatively harmless looking floes. The night was clear and stars visible. Chief Officer Murdock was in charge of the bridge.

The first intimation of the presence of the iceberg that he received was from the lookout in the crow's nest. They were so close upon the berg at this moment that it was practically impossible to avoid a collision with it.

The first officer did what other unstartled and alert commanders would have done under similar circumstances—that is, he made an effort by going full speed ahead on his starboard propeller and reversing his port propeller, simultaneously throwing his helm over, to make a rapid turn and clear the berg.

RIPPED BOTTOM OPEN

These maneuvers were not successful. He succeeded in preventing his bow from crashing into the ice cliff, but nearly the entire length of the great ship on the starboard side was ripped.

WRECK OF THE TITANIC

The speed of the *Titanic,* estimated to be at least twenty-one knots, was so terrific that the knifelike edge of the iceberg's spur protruding under the sea cut through her like a can opener.

The shock was almost imperceptible. The first officer did not apparently realize that the ship had received its death wound and none of the passengers it is believed had the slightest suspicion that anything more than a usual minor accident had happened. Hundreds who had gone to their berths and were asleep were not awakened by the vibration.

RETURNED TO CARD GAME

To illustrate the placidity with which practically all the men regarded the accident it was related that four who were in the smoking room playing bridge calmly got up from the table, and, after walking on deck and looking over the rail, returned to their game. One of them had left his cigar on the card table, and while the three others were gazing out on the sea he remarked that he couldn't afford to lose his smoke, returned for his cigar, and came out again.

The three remained only for a few moments on deck. They resumed their game under the impression that the ship had stopped for reasons best known to the commander and not involving any danger to her. The tendency of the whole ship's company except the men in the engine department, who were made aware of the danger by the inrushing water, was to make light of it and in some instances even to ridicule the thought of danger to so substantial a fabric.

SLOW TO REALIZE PERIL

Within a few minutes stewards and other members of the crew were sent around to arouse the people. Some utterly refused to get up. The stewards had almost to force the doors of the staterooms to make the somnolent appreciate their peril.

Mr. and Mrs. Astor were in their room and saw the ice vision flash by. They had not appreciably felt the gentle shock and supposed then nothing out of the ordinary had happened. They were both dressed and came on deck leisurely.

It was not until the ship began to take a heavy list to starboard that a tremor of fear pervaded it.

LAUNCHED BOATS SAFELY

The crew had been called to clear away the lifeboats, of which there were twenty, four of which were collapsible. The boats that were lowered on the port side of the ship touched the water without capsizing. Some of the others lowered to starboard, including one collapsible, were capsized. All hands on the collapsible boats that practically went to pieces were rescued by the other boats.

WRECK OF THE TITANIC

Sixteen boats in all got away safely. It was even then the general impression that the ship was all right and there is no doubt that that was the belief of even some of the officers.

At the lowering of the boats the officers superintending it were armed with revolvers, but there was no necessity for using them as there was nothing in the nature of a panic and no man made an effort to get into a boat while the women and children were being put aboard.

BEGAN TO JUMP INTO SEA

As the ship began to settle to starboard, heeling at an angle of nearly forty-five degrees, those who had believed it was all right to stick by the ship began to have doubt and a few jumped into the sea. These were followed immediately by others and in a few minutes there were scores swimming around. Nearly all of them wore life preservers.

One man who had a Pomeranian dog leaped overboard with it and striking a piece of wreckage was badly stunned. He recovered after a few minutes and swam toward one of the lifeboats and was taken aboard. Most of the men who were aboard the *Carpathia,* barring the members of the crew who had manned the boats, had jumped into the sea as the *Titanic* was settling.

Under instructions from officers and men in charge, the lifeboats were rowed a considerable distance from the ship itself in order to get away from the possible suction that would follow the foundering. The marvelous thing about the disappearance was so little suction as to be hardly appreciable from the point where the boats were floating.

There was ample time to launch all boats before the *Titantic* went down, as it was two hours and twenty minutes afloat.

So confident were all hands that it had not sustained a mortal wound that it was not until 12:15 a. m., or thirty-five minutes after the berg was encountered, that the boats were lowered. Hundreds of the crew and a large majority of the officers, including Capt. Smith, stuck to the ship to the last.

It was evident after there were several explosions, which doubtless were the boilers blowing up, that it had but a few minutes more of life.

SINKS WITH LITTLE FLURRY

The sinking ship made much less commotion than the horrified watchers in the lifeboats had expected. They were close enough to the broken vessel to see clearly the most grewsome details of the foundering. All the spectators agreed that the shattered sections of the ship went down so quietly as to excite wonder.

Some of the rescued were scantily clad and suffered exceedingly from the cold, but the majority of them

THE CRADLE IN WHICH THE S.S. TITANIC WAS BUILT

THE DESTRUCTION OF THE TITANIC.   THE MEETING OF THE TITANS.

WRECK OF THE TITANIC

were prepared for the emergency. In the darkness aboard the ship that came shortly after the collision it was impossible for those in the boats to distinguish the identity of any of the persons who leaped into the sea. It is believed that nearly all cabin passengers who had not gone overboard immediately after the boats were launched vanished with the officers and crew.

HAD TIME TO DRESS

Some of the stewards who formed part of the lifeboat crew say that after the ship hit the berg the majority of the cabin passengers went back to their staterooms and that it was necessary to rout them out and in some instances force life preservers upon them. All agree that the engines of the ship were stopped immediately after she had made the ineffectual turn to clear the berg.

The lifeboats' crews were made up of stewards, stokers, coal trimmers, and ordinary seamen. It is said that the davits were equipped with a new contrivance for the swift launching of the boats, but that the machinery was so complicated and the men so unfamiliar with it that they had trouble in managing it.

CHAPTER IV

THRILLING STORY OF THE WRECK

TOLD BY L. BEASLEY, M. A., OF CAMBRIDGE UNIVERSITY, ENGLAND.

"The voyage from Queenstown had been quite uneventful; very fine weather was experienced and the sea was quite calm. The wind had been westerly to southwesterly the whole way, but very cold, particularly the last day; in fact, after dinner on Sunday evening it was almost too cold to be out on deck at all.

"I had been in my berth for about ten minutes when at about 11:40 P. M. I felt a slight jar and then soon after a second one, but not sufficiently large to cause any anxiety to anyone however nervous they may have been. The engines stopped immediately afterward and my first thought was—'she has lost a propeller.'

"I went up on the top deck in a dressing gown, and found only a few people there, who had come up similarly to inquire why we had stopped, but there was no sort of anxiety in the minds of anyone.

"We saw through the smoking-room window a game of cards going on and went in to inquire if the players knew anything; it seems they felt more of the jar, and looking through the window had seen a huge iceberg go by close to the side of the boat. They thought we had

WRECK OF THE TITANIC

just grazed it with a glancing blow, and the engines had been stopped to see if any damage had been done. No one, of course, had any conception that she had been pierced below by part of the submerged iceberg.

"The game went on without any thought of disaster, and I retired to my cabin to read until we went on again. I never saw any of the players or the onlookers again. A little later, hearing people going upstairs, I went out again and found every one wanting to know why the engines had stopped.

"No doubt many were awakened from sleep by the sudden stopping of a vibration to which they had become accustomed during the four days we had been on board. Naturally, with such powerful engines as the *Titanic* carried, the vibration was very noticeable all the time, and the sudden stopping had something the same effect as the stopping of a loud-ticking grandfather's clock in a room.

"PUT ON LIFE BELTS"

"On going on deck again I saw that there was an undoubted list downward from stern to bow, but knowing of what had happened concluded some of the front compartments had filled and weighed her down. I went down again to put on warmer clothing, and as I dressed heard an order shouted:

" 'All passengers on deck with life belts on.'

"We walked slowly up with them tied on over our clothing, but even then presumed this was a wise precaution the captain was taking, and that we should return in a short time and retire to bed.

"There was a total absence of any panic or any expressions of alarm, and I suppose this can be accounted for by the exceedingly calm night and the absence of any signs of the accident.

REAL PERIL WAS HIDDEN

"The ship was absolutely still and except for a gentle tilt downward, which I do not think one person in ten would have noticed at that time, no signs of the approaching disaster were visible. She lay just as if she were waiting the order to go on again when some trifling matter had been adjusted. But in a few moments we saw the covers lifted from the boats and the crews allotted to them standing by and curling up the ropes which were to lower them by the pulley blocks into the water.

"We then began to realize it was more serious than had been supposed, and my first thought was to go down and get more clothing and some money, but seeing people pouring up the stairs decided it was better to cause no confusion to people coming up by doing so.

"Presently we heard the order:

" 'All men stand back away from the boats and all

ladies retire to next deck below'—the smoking-room deck or B deck. The men all stood away and remained in absolute silence, leaning against the end railings of the deck or pacing slowly up and down.

"The boats were swung out and lowered from A deck. When they were to the level of B deck, where all the ladies were collected, the ladies got in quietly, with the exception of some who refused to leave their husbands. In some cases they were torn from them and pushed into the boats, but in many instances they were allowed to remain because there was no one to insist they should go.

"Looking over the side, one saw boats from aft already in the water, slipping quietly away into the darkness, and presently the boats near to me were lowered and with much creaking as the new ropes slipped through the pulley blocks down the ninety feet which separated them from the water. An officer in uniform came up as one boat went down and shouted: 'When you are afloat, row round to the companion ladder and stand by with the other boats for orders.'

DISCIPLINE HOLDS GOOD

" 'Aye, aye, sir,' came up the reply, but I do not think any boat was able to obey the order. When they were afloat and had the oars at work the condition of the rapidly settling boat was so much more a sight for alarm for those in the boats than those on board that in common prudence the sailors saw they could do nothing but row from the sinking ship to save at any rate some lives. They no doubt anticipated that suction from such an enormous vessel would be more than usually dangerous to a crowded boat mostly filled with women.

"All this time there was no trace of any disorder, panic or rush to the boats, and no scenes of women sobbing hysterically, such as one generally pictures as happening at such times; every one seemed to realize so slowly that there was imminent danger.

"When it was realized that we might all be presently in the sea, with nothing but our life belts to support us until we were picked up by passing steamers, it was extraordinary how calm every one was and how completely self-controlled.

"One by one the boats were filled with women and children, lowered and rowed away into the night. Presently the word went round among the men, 'the men are to be put into the boats on the starboard side.' I was on the port side, and most of the men walked across the deck to see if this was so.

"I remained where I was, and presently heard the call:

"'Any more ladies?' Looking over the side of the ship, I saw the boat, No. 13, swinging level with B deck, half full of ladies.

"Again the call was repeated:

"'Any more ladies?'

"I saw none come on and then one of the crew looked up and said: 'Any ladies on your deck, sir?'

"'No,' I replied.

"'Then you had better jump.'

"I dropped in and fell in the bottom, as they cried 'lower away.' As the boat began to descend two ladies were pushed hurriedly through the crowd on B deck and heaved over into the boat, and a baby of 10 months passed down after them. Down we went, the crew calling to those lowering which end to keep her level. 'Aft,' 'stern,' 'both together,' until we were some ten feet from the water, and here occurred the only anxious moment we had during the whole of our experience from leaving the deck to reaching the *Carpathia*.

NEW PERIL THREATENED

"Immediately below our boat was the exhaust of the condensers, a huge stream of water pouring all the time from the ship's side just above the water line. It was plain we ought to be quite a way from this not to be swamped by it when we touched water. We had no officer aboard, nor petty officer or member of the crew to take charge. So one of the stokers shouted: 'Some one find the pin which releases the boat from the ropes and pull it up.' No one knew where it was. We felt as well as we could on the floor and sides, but found nothing, and it was hard to move among so many people—we had sixty or seventy on board.

"Down we went and presently floated with our ropes still holding us, the exhaust washing us away from the side of the vessel and the swell of the sea urging us back against the side again. The result of all these forces was an impetus which carried us parallel to the ship's side and directly under boat No. 14, which had filled rapidly with men and was coming down on us in a way that threatened to submerge our boat.

SOUND FAILED TO CARRY

"'Stop lowering 14,' our crew shouted, and the crew of No. 14, now only twenty feet above, shouted the same. But the distance to the top was some seventy feet and the creaking pulleys must have deadened all sound to those above, for down it came—fifteen feet, ten feet, five feet, and a stoker and I reached up and touched her swinging above our heads. The next drop would have brought it on our heads, but just before it dropped another stoker sprang to the ropes with his knife.

"'One,' I heard him say; 'two,' as his knife cut through the pulley ropes, and the next moment the exhaust steam had carried us clear, while boat 14 dropped into the space we had the moment before occupied, our gunwales almost touching.

"We drifted away easily as the oars were got out and headed directly away from the ship. The crew seemed to me to be mostly cooks in white jackets, two to an oar, with a stoker at the tiller.

"The captain-stoker told us that he had been on the sea twenty-six years and had never seen such a calm night on the Atlantic. As we rowed away from the *Titanic* we looked back from time to time to watch it, and a more striking spectacle it was not possible for any one to see.

TITANIC GREAT IN DEATH

"In the distance it looked an enormous length, its great bulk outlined in black against the starry sky, every porthole and saloon blazing with light. It was impossible to think anything could be wrong with such a leviathan were it not for that ominous tilt downward in the bow, where the water was by now up to the lowest row of portholes. Presently about 2 A. M., as near as I can remember, we observed it settling very rapidly, with the bow and bridge completely under water, and concluded it was now only a question of minutes before it went; and so it proved.

"It slowly tilted straight on end, with the stern vertically upward, and as it did, the lights in the cabins and saloons, which had not flickered for a moment since we left, died out, came on again for a single flash, and finally went altogether.

"To our amazement the *Titanic* remained in that upright position, bow down, for a time which I estimate as five minutes, while we watched at least 150 feet of the *Titanic* towering above the level of the sea and looming black against the sky. Then the ship dived beneath the waters.

HEARD CRY OF DYING

"And then, with all these, there fell on the ear the most appalling noise that human being ever listened to —the cries of hundreds of our fellow beings struggling in the icy cold water, crying for help with a cry that we knew could not be answered. We longed to return and pick up some of those swimming, but this would have meant swamping our boat and loss of life to all of us.

THE CARPATHIA APPEARS

"Our rescuer showed up in a few hours, and as it swung round we saw its cabins all alight and knew it must be a large steamer. It was now motionless, and we had to row to it. Just then day broke, a beautiful, quiet dawn with faint pink clouds just above the horizon, and a new moon whose crescent just touched the waters."

"The passengers, officers and crew gave up gladly their staterooms, clothing and comforts for our benefit, all honor to them."

WRECK OF THE TITANIC

The English Board of Trade passenger certificate on board the *Titanic* showed approximately 3,500. The same certificate called for lifeboat accommodation for approximately 950 in the following boats:

Fourteen large lifeboats, two smaller boats and four collapsible boats.

Life-preservers were accessible and apparently in sufficient number for all on board.

The approximate number of passengers carried at the time of the collision was:

First class, 330; second class, 320; third class, 750; total, 1,400. Officers and crew, 940. Total, 2,340.

Of the foregoing about the following were rescued by the steamship *Carpathia*:

First class, 210; second class, 125; third class, 200; officers, 4; seamen, 39; stewards, 96; firemen, 71; total, 210 of the crew. The total, about 745 saved, was about 80 per cent of the maximum capacity of the lifeboats.

CHAPTER V
RESCUE OF THE SURVIVORS

ONLY 745 OF THE 2,340 SOULS ABOARD DOOMED LINER SAVED BY THE LIFEBOATS—LITTLE SHOCK FELT WHEN THE ICEBERG WAS STRUCK BY THE TITANIC.

Freighted with its argosy of woe, disaster and death, bringing glad reunion to some, but misery unutterable to many, the *Carpathia,* with the survivors of the lost *Titanic* aboard, came back to a grief-stricken city and nation four days after the disaster. It was received by awe-stricken thousands whose conversation was conducted in whispers.

The story it brought home was one to crush the heart with its pathos, but at the same time to thrill it with pride in the manly and womanly fortitude displayed in the face of the most awful peril and inevitable death.

As the *Titanic* went down, according to the story of those who were among the last to leave the wounded hulk, the ship's band was playing.

ESTIMATED 1,595 DEAD

As brought to port by the *Carpathia,* the death list was placed at 1,601. The *Titanic* had aboard 2,340 persons, of whom 745 were picked up. Six of the latter succumbed to the exposure they had undergone before the *Carpathia* reached port.

Not only was the *Titanic* tearing through the April night to its doom with every ounce of steam crowded on, but it was under orders from the general officers of the line to make all the speed of which she was capable. This was the statement made by J. H. Moody, a quartermaster of the vessel and helmsman on the night of the disaster. He said the ship was making twenty-one knots an hour, and the officers were striving to live up to the orders to smash the record.

"It was close to midnight," said Moody, "and I was on the bridge with the second officer, who was in command. Suddenly he shouted, 'Port your helm!' I did so, but it was too late. We struck the submerged portion of the berg."

LITTLE SHOCK FELT

As nearly as most of the passengers could remember, the *Titanic,* sliding through the water at no more speed than had been consistently maintained during all of the trip, slid gracefully a few feet out of the water with just the slightest tremble. It rolled slightly; then it pitched. The shock, scarcely noticeable to those on board, drew a few loungers over to the railings. Officers and petty officers were hurrying about. There was no destruction within the ship, at least not in the sight of the passengers.

There was no panic. Everything that could be seen tended to alleviate what little fear had crept into the minds of the passengers, who were more apprehensive than the regular travelers who cross the ocean at this season of the year and who were more used to experiencing those small quivers.

Not one person aboard the *Titanic,* unless possibly it was the men of the crew, who were working far below, knew the extent of the injuries it had sustained. Many of the passengers had taken time to dress, so sure were they that there was no danger. They came on deck, looked the situation over and were unable to see the slightest sign that the *Titanic* had been torn open beneath the water line.

When the passengers' fear had been partly calmed and most of them had returned to their staterooms or to the card games in which they were engaged before the quiver was felt, there came surging through the first cabin quarters a report, that seemed to have drifted in from nowhere, that the ship was sinking.

How this word crept in from outside no one seems now to know. Immediately the crew began to man the boats.

Then came the shudder of the riven hulk of the once magnificent steamship as it receded from the shelving ice upon which it had driven, and its bow settled deeply into the water.

WRECK OF THE TITANIC

"We're lost! We're lost!" was the cry that rose from hundreds of throats. "The ship is sinking. We must drown like rats!"

Women in evening gowns, with jewels about their necks, knelt on deck, amid the vast, fear-stricken throng, crowded about the lifeboats and prayed for help. Others, clad in their nightclothing, begged the officers to let them enter the boats.

"Everybody to the boats!" was the startling cry that was repeated from end to end of the *Titanic*.

"Women and children first!" was the hoarse order that went along the line of lifeboats.

Without food, without clothing and with only the clothes in which they stood when the shock came, the women were tossed over the rails of the lifeboats, the davits were swung out, a few men were picked to man the oars, an officer to command the boat and the order to "lower away" was shouted. The little craft, laden with living freight, were launched.

NO CHOICE BETWEEN CLASSES

Men whose names and reputation were prominent in two hemispheres were shouldered out of the way by roughly dressed Slavs and Hungarians. Husbands were separated from their wives in the battle to reach the boats. Tearful leave-takings as the lifeboats, one after another, were filled with sobbing women and lowered upon the ice-covered surface of the ocean were heart-breaking.

There was no time to pick or choose. The first woman to step into a lifeboat held her place even though she were a maid or the wife of a Hungarian peasant. Many women clung to their husbands and refused to be separated. In some cases they dragged their husbands to the boats and in the confusion the men found places in the boats.

Before there was any indication of panic, Henry B. Harris, a theatrical manager of New York, stepped into a boat at the side of his wife before it was lowered.

"Women first!" shouted one of the ship's officers. Mr. Harris glanced up and saw that the remark was addressed to him.

"All right," he replied, coolly.

"Goodby, my dear," he said, as he kissed his wife, pressed her a moment to his breast and then climbed back to the *Titanic's* deck.

FLEET DREW AWAY

One by one the little fleet drew away from the towering sides of the giant steamship, whose decks were already reeling as it sank lower in the water.

"The *Titanic* is doomed!" was the verdict that passed from lip to lip.

"We will sink before help can come!"

Water poured into every compartment of the 800-

Photo Underwood & Underwood, N. Y.
HOISTING TITANIC LIFEBOAT FILLED WITH RESCUED ABOARD THE CARPATHIA

HAROLD BRIDE
A Titanic wireless operator, being carried ashore from Carpathia. He jumped into the sea and was rescued, but his feet were badly frozen.

WRECK OF THE TITANIC

foot hull, where great plates had been torn apart and huge rivets were sheared off as though they were so much cheese.

Pumps were started in the engine-room, but the water poured into the great hull in such torrents through scores of rents that all knew the fight to save the steamship was hopeless.

Overhead the wireless buzzed the news to the other steamships. The little fleet of lifeboats withdrew to a safe distance and the 1,595 left on board with no boats waited for the merciful death plunge which ended all.

WOMEN SAVED FIRST

A few spars, a box or two, a few small pieces of other wreckage, were the only portions of the *Titanic* corpse that lived on the water surface to be beheld by the persons on board the *Carpathia* when it rushed to the rescue. It was just breaking day as the rescue work was completed.

So exhausted were the survivors that scarcely any of them were able to tell their story of what actually had happened until late in the afternoon of Monday. It seemed impossible to obtain a complete story of the tragedy.

FEW INJURED ON WRECK

Certainly few of the Titanic passengers were hurt on board that great vessel. Few of the persons who came in among the survivors on the *Carpathia* bore any marks of injury. Their sufferings were caused chiefly by exposure, shock and grief. The latter was terrible. Many of the women had walked into a boat after kissing their husbands good-by.

The women in the lifeboats saw their loved ones plunge to death. The survivors' boats were bobbing along in the waves all within a radius of half a mile of the great *Titanic,* when, with a roar and burst of spray, it settled and passed out of sight for the last time.

Then began one of the most tortuous experiences for the helpless women in the drifting lifeboats that human beings ever were compelled to endure.

It was black night. Fortunately several of the men who were saved and some of the few petty officers who had aided in manning the lifeboats had a few matches in their pockets. Their torches were improvised from letters and scraps of papers that were found in their pockets. There was nothing to be seen.

SIGNALED WITH TORCHES

The torches, the only hope of those who thought they were doomed to death, were being carefully guarded and many times those who held them were implored to light them in the faint hope that rescue was closer at hand than even the most sanguine could have believed.

But the strong prevailed and it was not until the first rocket was seen to shoot heavenward from the *Carpathia* that the first of the torches was lighted and its filmy blaze shot up as high as was possible when one of the men, held on the shoulders of five others, stood up and waved the flaming papers until they burned down to his finger tips.

The desolate groups huddled together in the tossing and rolling tiny craft could not tell whether their torch had been seen by the ship that was firing the rockets. They waited fifteen minutes and the operation was repeated.

Then the huge bulk of the *Carpathia* took form in the gray of the breaking morning and it swept swiftly down into the center of a widely separated fleet of lifeboats with their human freight, then more dead than alive. They had been for approximately six hours in the open with the waves sending spray and at intervals whole barrelsful of water in upon them. They were drenched and the severe cold was freezing their clothing to their bodies. Only a few of them were able to walk when finally it came their turn to be taken on board the *Carpathia*.

The *Carpathia's* sailors went after those lying unconscious in the bottom of the lifeboats, lifted them up to other sailors standing on the *Carpathia's* ladders.

Everything that could be done for the survivors was done on the *Carpathia*.

Several of them had been cut and bruised in their attempts to get into the lifeboats and by falling from exhaustion during the awful ordeal they were compelled to pass through while waiting for the *Carpathia* to come to their relief. These were given surgical care. The others were placed in bed and few if any of them were able during the rest of the voyage to go on deck.

TELLS OF THE RESCUE

A passenger on the *Carpathia* made the following statement:

"I was awakened at about half past twelve at night by a commotion on the decks which seemed unusual, but there was no excitement. As the boat was moving I paid little attention to it, and went to sleep again. About three o'clock I again awakened. I noticed that the boat had stopped. I went to the deck. The *Carpathia* had changed its course.

"Lifeboats were sighted and began to arrive—and soon, one by one, they drew up to our side. There were sixteen in all, and the transferring of the passengers was most pitiable. The adults were assisted in climbing the rope ladder by ropes adjusted to their waists. Little children and babies were hoisted to the deck in bags.

FEW IN SOME BOATS

"Some of the boats were crowded, a few were not half full. This I could not understand. Some people were in full evening dress. Others were in their night clothes and were wrapped in blankets. These, with immigrants in all sorts of shapes, were hurried into the saloon indiscriminately for a hot breakfast. They had been in the open boats four and five hours in the most biting air I ever experienced.

"There were husbands without wives, wives without husbands, parents without children and children without parents. But there was no demonstration. No sobs— scarcely a word spoken. They seemed to be stunned. Immediately after breakfast, divine service was held in the saloon.

"One woman died in the lifeboat; three others died soon after reaching our deck. Their bodies were buried in the sea at five o'clock that afternoon. None of the rescued had any clothing except what they had on, and a relief committee was formed and our passengers contributed enough for their immediate needs.

TELLS OF FINAL PLUNGE

"When its lifeboats pushed away from the *Titanic,* the steamer was brilliantly lighted, the band was playing and the captain was standing on the bridge giving directions. The bow was well submerged and the keel rose high above the water. The next moment everything disappeared. The survivors were so close to the sinking steamer that they feared the lifeboats would be drawn into the vortex.

"On our way back to New York we steamed along the edge of a field of ice which seemed limitless. As far as the eye could see to the north there was no blue water. At one time I counted thirteen icebergs."

—*Cleveland Plain Dealer*

CHAPTER VI

SURVIVORS REACH NEW YORK

HOSPITALS SENT AMBULANCES AND NURSES—INVESTIGATION BY THE SENATE DECIDED UPON

At 8 o'clock automobiles and carriages containing relatives and friends of the survivors began arriving at the White Star pier. When the *Carpathia* was sighted coming up the river at 8:45, more than 500 automobiles and other vehicles were packed within the police lines.

Significant of the tragic side of the event was the frequent arrivals of ambulances and auto trucks from all the big department stores, filled with cots, invalid chairs and surgical appliances. Right of way was given the ambulances and they were permitted to park directly alongside the pier entrance.

HOSPITALS SENT NURSES

From St. Vincent's Hospital came twelve black-robed sisters to nurse the injured, and all the ambulances of the institution except one. The full surgical staff of the hospital also was in attendance. Ambulances and surgeons were on hand from St. Luke's Hospital, Bellevue, Roosevelt and Flower hospitals, and a great number of physicians who had volunteered their services.

The Sisters of Charity found work to do before the arrival of the *Carpathia*. Women in the throng awaiting relatives became hysterical with dread and anxiety and the black-robed sisters went to them, put their arms about them and comforted them and administered restoratives.

Eva Booth, commander of the Salvation Army, and fifty assistants, who meet all incoming vessels to minister to immigrants, were allowed within the police lines, but they were turned back at the entrance of the Cunard pier and only Miss Booth and three of her party were admitted.

BROKERS BROUGHT $20,000

Among those on the pier were six members of the New York Stock Exchange, with $20,000, which had been collected on the floor of the exchange. They had instructions to use the money among the steerage passengers in any way they saw fit.

The women of the relief committee to look after the steerage passengers arrived in autos and theater buses, in which the sufferers were to be taken to hospitals or shelters. Gimbel Brothers sent all their delivery wagons to the pier, laden with first aid appliances and cots, and placed them at the disposal of the women's relief committee. In addition, the firm announced they would provide quarters for 200 sufferers overnight in their store.

WRECK OF THE TITANIC

CALLED FOR MORE NURSES

Relatives and friends of the survivors had reached the pier before half past eight o'clock, but for another half hour automobiles arrived containing physicians and nurses and loaded with first aid appliances. The surgeons and nurses were in working attire, the women in white gowns and caps, the surgeons in white duck trousers and jackets.

A party of four surgeons and ten nurses arrived in three automobile buses and as they hurried to the pier one of them said they had been sent by Mrs. William K. Vanderbilt.

In spite of the number of physicians that had reached the pier at 8:30, it was found there was a dearth of nurses and hurried calls were sent out to all the city institutions and private hospitals and nurses' exchanges. In response to these calls nurses began arriving in taxicabs and autos, and before the *Carpathia* was warped into its pier there were more than 200 nurses awaiting to go on board.

Ropes dotted with green lights were stretched for seventy-five yards in front of the piers to hold back the throngs. No one without a special permit was allowed beyond these ropes.

The Pennsylvania Railroad Company had a special train waiting at its station at Thirty-fourth street and a number of taxicabs to convey survivors desiring to go to Philadelphia to their friends.

News that the *Carpathia* was outside of the harbor and rapidly approaching sent thousands of persons to vantage points along the city's water front. At the Battery, the first point on Manhattan Island which the rescue ship would pass, a crowd estimated at 10,000 persons assembled. Other vantage points further uptown were crowded with spectators eager to catch the first glimpse of the approaching *Carpathia*.

INVESTIGATION DECIDED ON

Senator William Alden Smith of Michigan and Senator Newlands of Nevada arrived in New York at 9 p. m. April 18 to summon survivors of the *Titanic* and officials of the International Mercantile Marine to testify before the Senate subcommittee appointed to investigate the disaster of the sea.

When the senators arrived at the Pennsylvania station they were informed that the *Carpathia* was at its pier. The committee had intended boarding a revenue cutter and going down the bay to meet the *Carpathia* and boarding it. Upon learning this the senators secured cabs and announced they were going direct to the pier.

CHAPTER VII

LAST MAN OFF TELLS HORRORS OF SHIPWRECK

COLONEL GRACIE, U. S. A., RESCUED AFTER GOING DOWN ON TITANIC'S TOPMOST DECK—HEROES ON ALL SIDES—MRS. ISIDOR STRAUS DROWNED, REFUSING TO DESERT HUSBAND—ASTOR PRAISED FOR CONDUCT.

Colonel Archibald Gracie, U. S. A., the last man saved after the wreck of the *Titanic,* went down with the vessel, but was picked up. He was met at the dock in New York by his daughter, who had arrived from Washington, and his son-in-law, Paul H. Fabricius.

Colonel Gracie told a remarkable story of personal hardship and denied emphatically reports that there was any panic on board the steamship after the disaster. He praised in the highest terms the behavior of both the passengers and crew and paid a high tribute to the heroism of the women passengers.

"Mrs. Isidor Straus," said Colonel Gracie, "went to her death because she would not desert her husband. Although he pleaded with her to take her place in the boat, she steadfastly refused, and when the ship settled at the head the two were engulfed by the wave that swept the vessel."

DRIVEN TO TOP DECK

Colonel Gracie told how he was driven to the topmost deck when the ship settled and was the sole survivor after the wave that swept it just before its final plunge had passed.

"I jumped with the wave," said he, "just as I often have jumped with the breakers at the seashore. By great good fortune I managed to grasp the brass railing on the deck above, and I hung on by might and main.

"When the ship plunged down I was forced to let go and was swirled around and around for what seemed to be an interminable time. Eventually I came to the surface to find the sea a mass of tangled wreckage.

"Luckily, I was unhurt, and, casting about, managed to seize a wooden grating floating near by. When I had recovered my breath I discovered a canvas and cork life raft which had floated up.

THIRTY SAVED ON RAFT

"A man whose name I did not learn was struggling toward this rafte from some wreckage to which he had clung. I cast off and helped him to get onto the raft, and we then began the work of rescuing those who had jumped into the sea and were floundering in the water.

"When dawn broke there were thirty of us on the raft, standing knee-deep in the icy water and afraid to move lest the craft be overturned.

"Several other unfortunates, benumbed and half dead, besought us to save them, and one or two made efforts to reach us, but we had to warn them away. Had we made any effort to save them we all might have perished.

LONG HOURS OF HORROR

"The hours that elapsed before we were picked up by the *Carpathia* were the longest and most terrible that I ever spent. Practically without any sensation of feeling because of the icy water, we were almost dropping from fatigue.

"We were afraid to turn around to learn whether we were seen by passing craft, and when some one who was facing astern passed the word that something that looked like a steamer was coming up one of them became hysterical under the strain. The rest of us, too, were nearing the breaking point."

Colonel Gracie denied with emphasis that any men were fired upon, and declared that only once was a revolver discharged.

"This," the colonel said, "was done for the purpose of intimidating some steerage passengers who had tumbled into a boat before it was prepared for launching. The shot was fired in the air, and when the foreigners were told that the next would be directed at them they promptly returned to the deck. There was no confusion and no panic."

Contrary to the general expectation, there was no jarring impact when the vessel struck, according to the army officer. He was in his berth when the *Titanic* smashed into the submerged portion of the iceberg and was aroused by the jar.

STOPPED WATCH FIXED TIME

Colonel Gracie looked at his watch, he said, and found it was just midnight. The ship sank with him at 2:22 a. m., for his watch stopped at that hour.

"Before I retired," said Colonel Gracie, "I had a long chat with Charles M. Hays, president of the Grand Trunk Railroad. One of the last things Mr. Hays said was this:

" 'The White Star, the Cunard and the Hamburg-American lines are devoting their attention and ingenuity to vying with one another to attain supremacy in luxurious ships and in making speed records. The time will soon come when this will be checked by some appalling disaster.'

"Poor fellow—a few hours later he was dead."

GAVE PRAISE TO ASTOR

"The conduct of Colonel John Jacob Astor was deserving of the highest praise," Colonel Gracie declared. "The millionaire New Yorker devoted all his energies to saving his young bride, formerly Miss Force of New York, who was in delicate health.

"Colonel Astor helped us in our efforts to get Mrs. Astor in the boat," said Colonel Gracie. "I lifted her into the boat and as she took her place Colonel Astor requested permission of the second officer to go with her for her own protection.

" 'No, sir,' replied the officer, 'not a man shall go on a boat until the women are all off.'

COLONEL AIDED WITH BOATS

"Colonel Astor then inquired the number of the boat which was being lowered away and turned to the work of clearing the other boats and reassuring the frightened and nervous women.

"By this time the ship had begun to list frightfully to port. This became so dangerous that the second officer ordered every one to rush to starboard.

"This we did, and we found the crew trying to get a boat off in that quarter. Here I saw the last of John B. Thayer and George B. Widener of Philadelphia."

IGNORED WARNINGS CHARGED

Colonel Gracie said that despite the warnings of icebergs no slowing down of speed was ordered by the commander of the *Titanic*. There were other warnings, too, he said.

"In the twenty-four hours' run ending the 14th," declared Colonel Gracie, "the ship's run was 546 miles, and we were told that the next twenty-four hours would see even a better record posted.

"No diminution of speed was indicated in the run and the engines kept up their steady work. When Sunday evening came we all noticed the increased cold, which gave plain warning that the ship was in close proximity to icebergs or ice fields.

"The officers, I am credibly informed, had been advised by wireless from other ships of the presence of icebergs and dangerous floes in that vicinity. The sea was as smooth as glass and the weather clear, so that it seemed that there was no occasion for fear.

"When the vessel struck the passengers were so little alarmed that they joked over the matter. The few who appeared upon deck early had taken their time to dress properly and there was not the slightest indication of panic.

Photo Underwood & Underwood

J. BRUCE ISMAY
White Star Line Manager

SINKING A DERELICT—DISPOSING OF A MENACE TO NAVIGATION

"Some fragments of ice had fallen on the deck and these were picked up and passed around by some of the facetious ones, who offered them as mementos of the occasion. On the port side, a glance over the side failed to show any evidence of damage, and the vessel seemed to be on an even keel.

"James Clinch Smith and I, however, soon found the vessel was listing heavily. A few minutes later the officers ordered men and women to don life-preservers."

WOMEN REFUSED RESCUE

One of the last women seen by Colonel Gracie, he said, was Miss Evans, of New York, who virtually refused to be rescued, because "she had been told by a fortune teller in London that she would meet her death on the water."

A young English woman who requested that her name be omitted told a thrilling story of her experience in one of the collapsible boats, which was manned by eight of the crew from the *Titanic*. The boat was in command of the fifth officer, H. Lowe, whom she credited with saving the lives of many persons.

Before the lifeboat was launched Lowe passed along the port deck of the steamer, commanding the people not to jump into the boats and otherwise restraining them from swamping the craft. When the collapsible was launched Lowe succeeded in putting up a mast and a small sail.

The officer collected the other boats together, and, in case where some were short of adequate crews, directed an exchange by which each was adequately manned. He threw lines which linked the boats two by two, and all thus moved together.

Later on Lowe went back to the wreck with the crew of one of the boats and succeeded in picking up some of those who had jumped overboard, and were swimming about. On his way back to the *Carpathia* he passed one of the collapsible boats which was on the point of sinking with thirty passengers aboard, most of them in scant night clothing. They were rescued just in the nick of time.

CHAPTER VIII

HEROISM ON THE *TITANIC*

PRESIDENT TAFT'S ESTIMATE OF MAJOR BUTT—BEN GUGGENHEIM NOT A COWARD—HEROIC MUSICIANS —"NEARER, MY GOD, TO THEE."

When President Taft heard that women and children had perished in the wreck of the *Titanic* he spoke his estimate of Archie Butt in saying: "I do not expect, I do not want, to see him back." That Mr. Taft knew his man was proved by the words of the rescued.

Note this: Benjamin Guggenheim sent word to his wife: "Tell her I played the game out straight to the end. No woman shall be left aboard this ship because Ben Guggenheim was a coward."

And this: "And then Mrs. Straus would call him (Mr. Straus) by his first name and say her place was with him, that she had lived with him and that she would die with him." And Mr. Straus said: "I am not too old to sacrifice myself for a woman."

And this of Mrs. Allison: "The boat was full and she grasped Lorraine with one arm and her husband with the other and stood smiling as she saw us rowing away."

And this of Captain Smith: "He swam to where a baby was drowning, carried it in his arms to a lifeboat, and then swam back to his ship to die." And this, the command given by Captain Smith bringing order out of chaos: "Be British, my men."

And lastly: Kraus, Hume, Taylor, Woodward, Clark, Brailey, Breicoux and Hartley, when the last faint hope was gone, lined up on deck, stood in water up to their knees and played "Nearer, My God to Thee," as 1,500 souls passed from life.

HEROIC MUSICIANS

Except in the case of the English ship *Birkenhead,* when the soldiers on board stood at parade after the women and children had been taken into the boats and the band played the national air as the ship went down, we do not recall a parallel to the conduct of the musicians on board the *Titanic,* who, as all accounts agree, ceased not their inspiring ministrations until they were engulfed by the waves.

Indeed, it seems even to be a question if the later instance of heroism was not greater than the former, for the bandsmen on the *Birkenhead* were enlisted men, obeying orders like soldiers, while it is scarcely to be thought that the obligations of the musicians on the *Titanic* required them to play with death confronting them. There has been a marvelous upwelling of sympathy for the families made destitute by the awful catastrophe, and, perhaps, a too great multiplicity of relief funds; but there is, nevertheless, something especially appealing in Dr. Frank Damrosch's suggestion that a special contribution be asked for the families of those who gave courage and comfort to the doomed victims of the steamship; and died to do it.

MAJOR BUTT DIED LIKE A SOLDIER

A graphic story of the heroism of Major Archibald W. Butt on the *Titanic* was told feelingly by Miss Marie Young, a former resident of New York, before going to her home in Washington, D. C. Miss Young is believed to have been the last woman to leave the *Titanic* and the last of the survivors to have talked with the President's military aid. She and Major Butt had long been friends, Miss Young having been a special music instructor to the children of Theodore Roosevelt when he was President. Miss Young said:

"The last person to whom I spoke on board the *Titanic* was Archie Butt, and his good, brave face smiling at me from the deck of the steamer was the last I could distinguish as the lifeboat I was in pulled away from the steamer's side.

"Archie himself put me into the boat, wrapped blankets around me and tucked me in as carefully as if we were starting on a motor ride. He himself entered the boat with me, performing the little courtesies as calmly and with as smiling a face as if death was far away instead of being but a few moments removed from him. When he had carefully wrapped me up he stepped on the gunwale of the boat, and, lifting his hat, smiled down at me.

" 'Goodby, Miss Young,' he said, bravely and smilingly. 'Luck is with you. Will you kindly remember me to all the folks back home?

"Then he stepped to the deck of the steamer, and the boat I was in was lowered to the water. It was the last boat to leave the ship; of this I am perfectly certain. And I know that I am the last of those who were saved to whom Archie Butt spoke. As our boat was lowered and left the side of the steamer Archie was still standing at the rail looking down at me. His hat was raised, and the same old, genial, brave smile was on his face. The picture he made as he stood there, hat in hand, brave and smiling, was one that will always linger in my memory."

Mrs. Henry B. Harris, in an interview, also described the heroism of Major Butt. She said:

"Archie Butt was a major to the last. God never made a finer nobleman than he. The sight of that man, calm, gentle, and yet as firm as a rock, never will leave me. The American Army is honored by him, and the way he showed some of the other men how to behave

WRECK OF THE TITANIC

when women and children were suffering that awful mental fear that came when we had to be huddled in those boats. Major Butt was near me, and I know very nearly everything he did.

"When the order came to take to the boats he became as one in supreme command. You would have thought he was at a White House reception, so cool and calm was he. When the time came he was a man to be feared. In one of the earlier boats fifty women, it seemed, were about to be lowered, when a man, suddenly panic-stricken, ran to the stern of it. Major Butt shot one arm out, caught him by the neck, and jerked him backward like a pillow. His head cracked against a rail and he was stunned.

" 'Sorry,' said Major Butt; 'but women will be attended to first or I'll break every damned bone in your body.'

"The boats were lowered away one by one, and as I stood by my husband he said to me, 'Thank God for Archie Butt.' Perhaps Major Butt heard it, for he turned his face toward us for a second. Just at that time a young man was arguing to get into a lifeboat, and Butt had hold of the lad by the arm like a big brother and appeared to be telling him to keep his head.

"How inspiring he was. I stayed until almost the last and know what a man Archie Butt was. They put me in a collapsible boat. I was one of three women in the first cabin in the thing; the rest were steerage people. Major Butt helped those poor, frightened steerage people so wonderfully, tenderly and yet with such cool and manly firmness. He was a soldier to the last. He was one of God's greatest noblemen, and I think I can say he was an example of bravery even to the officers of the ship. He gave up his life to save others."

THE ETERNAL COLLISION

CHAPTER IX

THRILLING EXPERIENCES OF SURVIVORS

MARVELOUS BEHAVIOR OF MEN PASSENGERS—A SWEDISH OFFICER'S STORY—DISCIPLINE MAINTAINED TO THE END

FIRST WOMAN IN LIFEBOATS

Mrs. Dickinson Bishop, of Detroit, said:

"I was the first woman in the first boat. I was in the boat four hours before being picked up by the *Carpathia*. I was in bed at the time the crash came, got up and dressed and went back to bed, being assured there was no danger. There were very few passengers on the deck when I reached there. There was little or no panic, and the discipline of the *Titanic's crew was perfect*. Thank God my husband was saved also."

P. D. Daly of England said he was above deck A and that he was the last man to scramble into the collapsible boat. He said that for six hours he was wet to his waist with the icy waters that filled the boat nearly to the gunwales.

MEN PRAISED BY WOMAN

One of the few women able to give an account of the disaster was Miss Cornelia Andrews of Hudson,

N. Y. Miss Andrews said she was in the last boat to be picked up.

"The behavior of the men," she said, "was wonderful—the most marvelous I have ever beheld."

"Did you see any shooting?" she was asked.

"No," she replied, "but one officer did say he would shoot some of the steerage who were trying to crowd into the boats. Many jumped from the decks. I saw a boat sink."

Miss Andrews was probably referring to the collapsible boat which overturned. She said that the sinking of the ship was attended by a noise such as might be made by the boilers exploding. She was watching the ship, she said, and it looked as if it blew up.

STORY BY SWEDISH OFFICER

Lieutenant Hakan Bjornstern Steffanson of the Swedish army, who was journeying to this country on the *Titanic* to see about the exportation of pulp to Sweden, narrowly escaped being carried down in the sinking ship when he leaped out from a lower deck to a lifeboat that was being lowered past him. Henry Woolner of London also made the leap in safety. Lieutenant Steffanson thinks he made the last boat to leave the ship and was only about a hundred yards away when it went down with a sudden lurch.

He told about his experience as he lay in the bed at the Hotel Gotham, utterly worn out by the strain he had been under despite his six feet of muscle. It was also the first time he had discarded the dress suit he had worn since the shock of collision startled him from his chair in the cafe where he and Mr. Woolner were talking.

"It was not a severe shock," said the lieutenant. "It did not throw any one from his seat; rather it was a twisting motion that shook the boat terribly. Most of the women were in bed. We ran up to the smoking room, where most of the men were rushing about trying to find out what was the matter, but there was a singular absence of apprehension, probably because we believed so thoroughly in the massive hulk in which we were traveling.

SOUGHT TO CALM WOMEN

"We helped to calm some of the women and advised them to dress and then set about getting them in boats. There seemed to be really no reason for it, but it was done because it was the safest thing to do.

"The men went about their task quietly. Why should they have done otherwise—the shock was so slight to cause such ruin. Mr. Woolner and I then went to a lower deck. It was deserted, but as we wished to find out what had happened we went down a deck lower. Then for the first time did we realize the seriousness of that twisting which had rent the ship nearly asunder.

We saw the water pouring into the hull and where we finally stood water rose to our knees.

"Woolner and I decided to get out as quickly as we could and as we turned to rush upward we saw sliding down the portside of the drowning ship a collapsible lifeboat. Most of those it contained were from the steerage, but two of the women were from the first cabin. It was in charge of two sailors.

JUMPED INTO SWAYING BOAT

"'Let's not take any chances,' I shouted to Woolner, and as it came nearly opposite us, swinging in and out slowly, we jumped and fortunately landed in it. The boat teetered a bit and then swiftly shot down to the water. Woolner and I took oars and started to pull with all our might to get from the ship before she sank, for now there was little doubt of what would happen.

"We could see some gathered in the steerage, huddled together, as we pulled away, and then cries of fear came to us.

"We hardly reached a point a hundred yards away —and I believe the boat I was in was the last to get safely away—when the horrible screams came through the night and the ship plunged swiftly down. It was so terribly sudden, and then there was a vast quiet, during which we shivered over the oars and the women cried hysterically. Some of them tried to jump over-

WRECK OF THE TITANIC 77

board and we had to struggle in the shaky boat to hold them until they quieted down.

VICTIMS FLOATED TO SURFACE

"There was little widespread suction from the sinking ship, strange to say, and shortly after it went down people came to the surface, some of them struggling and fighting to remain afloat, and some were very still. But they all sank before we could reach them.

"It was bitterly cold and most of us were partly wet. It seemed hours before the *Carpathia* came up and took us aboard. Why, it was so cold that on board the *Titanic* we had been drinking hot drinks as if it were winter. The weather was absolutely clear, there was not the slightest fog or mist."

BOILER BLAST SPLIT VESSEL

Mrs. E. W. Carter left the *Carpathia* terribly shaken by her experience. She was met at the pier by Albert B. Ashforth. Mrs. Carter could not talk of the collision and the wreck, but Mr. Ashforth said that what had impressed her was the last boiler explosion.

"Mrs. Carter said that the shock of the collision was nothing," said Mr. Ashforth, "but the last boiler explosion tore the ship to pieces. She was in the last boat off."

What impressed E. Z. Taylor of Philadelphia most was the lack of excitement when the ship struck. He said he was on deck when the *Titanic* hit the iceberg and that he did not see any iceberg and did not think that anybody else did. Mr. Taylor said that he saw Mr. Ismay get into a boat fifteen minutes before the *Titanic* sank.

BAND PLAYED UNTIL END

Three sisters, Mrs. Robert C. Cornell, the wife of Magistrate Cornell; Mrs. E. B. Appleton, and Mrs. John Murray Brown of Acton, Mass., went immediately to the home of Magistrate Cornell and related to George S. Keyes, a son-in-law of Mrs. Brown, what they had gone through. Mrs. Brown's story is the most vivid, as she left the *Titanic* in the last boat that got away safely.

"The discipline was magnificent," she said. "The band played, marching from deck to deck, and as the ship was engulfed you could hear the music plainly. The last I saw of the band the musicians were up to their knees in water.

"My two sisters and I were separated and each got in different boats. The captain stood on the bridge, and when the water covered the ship he was offered assistance and told to get in one of the lifeboats, but he refused to do so.

WATCHED PARTING OF ASTORS

"Mrs. Astor was in the lifeboat with my sister, Mrs. Cornell. I saw Col. Astor help her into the boat. He said he would wait for the men. I saw him on the ship after our boat left the *Titanic.*

"We had a rough experience, many of the women having to use the oars. Mrs. Appleton's hands were badly torn, but I understand there was not a single case of illness among the survivors because of exposure.

"Picture a situation such as this! Another woman and myself were waiting to be helped into the lifeboat. The woman held my arm. I do not know her name. There was just one seat left in the boat. The woman said to the men, 'This woman has children; let her go first. I'll take the next boat.' I believe she was put in the next boat. That boat was swamped."

DROVE SEVERAL MEN BACK

Mrs. Ada Clark, an English woman who lost her husband in the wreck, stayed in her berth for half an hour after the collison.

"The shock was so light that it did not disturb me," she said, "and my husband told me to go back to sleep again. Then the stewardess came along and yelled, 'Everybody on deck.' There was no disturbance in filling the small boats. My husband put me in, kissed me goodbye and commended me to God. After I got into the boat two men tried to step in. An officer said that the boat was only for women and they stepped right back.

"I was in my night dress. The cold reached my brain and everybody in the boat was so benumbed from cold that we could not realize what a terrible thing had happened. Then somebody said, 'It's gone,' and we sat there without showing any emotion."

SAVED WITH HER CHILDREN

Mrs. Allen O. Becker, who is attached to the American Lutheran Missionary Society of Foreign Missions, and her three children, Ruth, 11; Marion, 8, and Richard, 6, were rescued from the *Titanic*.

She said she was awakened about 10:30 and a steward told her that everything was safe and that she could go back to sleep. In a half hour she was awakened by a steward who told her to take her three children in a hurry, as they were going to be put into a lifeboat. They did not get a chance to dress.

Mrs. Becker said that a steward took two of the children and she went with Ruth, but they all met in the same lifeboat. She said that they were in the boat until almost 5 o'clock when they were picked up.

JUMPED INTO SMALL LIFEBOAT

Abraham Hyman, a steerage passenger from Manchester, England, won his safety by leaving the steerage and going into the first cabin.

"I got alongside of a boat," he said, "and as they lowered it, full of passengers, I just crowded in beside the

Copyright Underwood & Underwood

MR. C. M. HAYS
President of the Grand Trunk Railroad, who lost his life. Mrs. Hays and daughter Margaret were saved

Photo Wm. L. Koehne
MRS. IDA HIPPACH AND DAUGHTER JEAN
Both of whom were rescued from the Titanic

man at the tiller. They could have taken fifteen more people in our boat. There was no commotion in the first cabin. I heard that a man was shot in a panic in the steerage. When our boat got into the water it drifted under the exhaust of the *Titanic* and we were nearly swamped. We rowed off for about half a mile and then saw the lights on the *Titanic* sink gradually out of sight. As the *Titanic* sank the lights went down, one after another."

Hyman said he heard of one man who had been sitting on a pile of deck chairs when the last explosion came who was blown off with the deck chairs. The man was found in the ocean on the deck chairs.

BOILERS REND GREAT SHIP

John Snyder and his wife of Indianapolis told how the boilers of the *Titanic* exploded and literally tore the ship to pieces.

"We were in our stateroom and I was asleep," Mr. Snyder said. "The jar that came when the ship struck the berg did not even awaken me, and later when my wife aroused me we could hear persons running about the ship. Then a steward came and told us that there was danger and that we had better dress at once.

"We did dress and went on the second deck. There seemed no great excitement among the passengers, although the officers of the ship were giving orders to the crew to lower the lifeboats and were telling the passengers to get into them.

"We were told to get into a boat and we did, although at the time I much preferred staying on the *Titanic*. It looked safe on the *Titanic* and far from safe in the lifeboat. Before we knew what was being done with us we were swung from the *Titanic* into the sea and then the boat was so crowded that the women lay on the bottom to give the crew a chance to row.

TITANIC SANK GRADUALLY

"We went about 200 yards from the *Titanic*. We could see nothing wrong except that the big boat seemed to be settling at the bow. Still we could not make ourselves believe that the *Titanic* would sink. But the *Titanic* continued to settle, and we could see the passengers plunging about the decks and hear their cries.

"We moved farther away. Suddenly there came two sharp explosions as the water rushed into the boiler room and the boilers exploded.

"The explosions counteracted the effect of the suction made when the big boat went to the bottom and it is more than probable that this saved some of the lifeboats from being drawn to the bottom.

EXPLOSIONS KILLED MANY

"Following the explosion we could see persons hanging to the side railings of the sinking boat. It is my opinion that many persons were killed by these explosions and were not drowned.

"Others of the passengers were tossed into the water. For an hour after the explosions we could see them swimming about in the water or floating on the life belts. We could hear their groans and their cries for help, but we did not go to them. To have done this would have swamped our own boat and everybody would have been lost. Several persons did float up to our boat and we took them on board.

"After we had got aboard the *Carpathia* we did not see J. Bruce Ismay until today, when he came on deck for a short time. He seemed badly broken up. You would hardly have known him."

PERIL UNKNOWN AT FIRST

A Mr. Chambers, one of the survivors, had this to say:

"The *Titanic* struck the iceberg. The passengers ran out, but, believing that the ship could not sink and being assured of this by the officers, again went back to their staterooms. After about two hours the alarm was sent out and the passengers started to enter the lifeboats. There was nothing like a panic at first, as all believed that there were plenty of lifeboats to go around."

After the lifeboat in which Mr. Chambers was had

gone about 400 yards from the ship, those in it saw the *Titanic* begin to settle quickly and there was a rush for the remaining lifeboats. One was swamped.

The great ship sank slowly by its head and no suction was felt by the boat in which Mr. Chambers was.

GREEN LANTERNS SAVED MANY

Henry Stengel of Newark said it was only the forethought of a member of a boat crew who was quick witted enough to snatch up three green lights that saved a number of the lives of those adrift in the tiny lifeboat.

"These green lights," he said, "shining through the darkness enabled the other boats' crews to keep close together in the ice filled waters."

Mr. Stengel put his wife in a boat and then followed. He said that early the next morning, shortly after they had been picked up, they saw floating far away a gigantic iceberg, with two peaks shining in the morning sun. There was the berg that sent the *Titanic* to the bottom, he thought.

JUMPED INTO SEA; PICKED UP

E. Z. Taylor of Philadelphia, one of the survivors, jumped into the sea just three minutes before the boat sank. He told a graphic story as he came from the *Carpathia*.

"I was eating when the boat struck the iceberg," he said. "There was an awful shock that made the boat tremble from stem to stern. I did not realize for some time what had happened. No one seemed to know the extent of the accident. We were told that an iceberg had been struck by the ship.

"I felt the boat rise and it seemed to me that it was riding over the ice. I ran out on deck and then I could see ice. It was a veritable sea of ice and the boat was rocking over it. I should say that parts of the iceberg were eighty feet high, but it had been broken into sections, probably by our ship.

"I jumped into the ocean and was picked up by one of the boats. I never expected to see land again. I waited on board the boat until the lights went out. It seemed to me that the discipline on board was wonderful."

SCENE AT RESCUE DESCRIBED

A passenger aboard the rescue ship *Carpathia,* Miss Sue Eva Rule, a sister of Judge Virgil Rule of St. Louis, Mo., detailed the thrilling scenes which marked the rescue of the survivors of the greatest maritime tragedy of the age.

"Unknown to the sleeping passengers, the ship turned abruptly to the north. None knew of the sudden change of course and the first intimation anybody got of the fact that anything unusual was about to take place was the order given the steward to prepare breakfast for 3,000.

"The tidings ran through the ship like wildfire and long before the Cunarder had come within the scene of the tragedy we were all on deck.

FIRST OF BOATS SIGHTED

"Just as day broke a tiny craft was sighted rowing towards us and as it came closer we saw women huddled together, the stronger ones manning the oars. The first to come aboard was a nurse maid who had wrapped in a coat an eleven-months-old baby, the only one of a family of five persons to be rescued.

"The men and women both seemed dazed. Most of them had almost perished with the cold, and some of them who had been literally thrown into the lifeboats perished from exposure.

"One of the most harrowing scenes I ever saw was the service of thanksgiving, followed by the prayers for the dead, which during the incoming of the little band of survivors, took place in the dining saloon of the *Carpathia*. The moans of the women and the cries of little children as their loss was brought home to them were heartrending. The hope that by some means their beloved ones would be saved never left the survivors.

SURVIVORS IN STRANGE DRESS

"How those who were saved survived the exposure is a miracle. One woman came aboard devoid of underwear, a Turkish towel wrapped about her waist served as a corset, while a magnificent evening wrap was her only protection.

"Women in evening frocks and white satin slippers and children wrapped in steamer rugs were ordinary sights and very soon the passengers themselves were almost in as bad a plight as the rescued. Trunks were unpacked and clothing distributed right and left. Finally the steamer rugs were ripped apart and sewed into impromptu garments.

"My first view of the first boat sighted led me to think we were picking up the crew of a dirigible. Back of the boat loomed in the shadowy dawn the huge iceberg which had sent the *Titanic* to the bottom. The lifeboat looked like the usual boat which swings from a balloon.

WOMEN DISCUSSED SCENES

"After an hour or so of rest the only relief the women who had been literally torn from their husbands seemed to have was in discussing the last scenes. Shooting was heard by many in the lifeboats just before the ship took its final plunge and sank from sight, and the opinion of many was that the men rather than drown shot themselves.

"Mrs. Astor, who was one of the first to come aboard, was taken at once to the captain's room. Others were distributed among the cabins, the *Carpathia's* passengers sleeping on the floors of the saloons, in the bathrooms, and on the tables throughout the ship in order to let the survivors of the wreck have as much comfort as the ship afforded.

"One woman came aboard with a six months' baby she had never seen until the moment it was thrust into her arms as she swung into the lifeboat. Nothing could equal the generosity and helpfulness of the *Carpathia's* passengers."

DOUBTED WORD AT FIRST

Mrs. Louise Mansfield Ogden, of Manhattan, described tonight how she felt when she heard the *Carpathia's* whistle sounding early in the morning. Mrs. Ogden asked her husband if there was a fog. Mr. Ogden had left the stateroom, however, and did not explain until some ten minutes later. The ship had then slowed down perceptibly, and Mrs. Ogden was pretty nervous.

Then her husband returned and told her that there had been a great accident and that the *Carpathia* was going to help.

"The passengers are asked to keep to their rooms," he said. "There isn't any need of being frightened. There's been no fire on our boat, but there has been an accident to the *Titanic*."

Mrs. Ogden thought that an accident to the *Titanic* was quite too ridiculous to think of and in that she shared the impression which, so she learned afterward, was current upon the *Titanic* after the latter had struck. Mrs. Ogden dressed hastily and went out on deck.

BOATS FILLED WITH SURVIVORS

"I saw there on the bosom of the ocean," she said, "a boat full of women and children. I suppose there must have been sailors there too, but I did not see them. There were only one or two women in evening dress, but most of them were clad in fur coats over their kimonos or nightgowns. They had on their evening slippers and silk stockings. Some of them wore hats.

"Far in the distance were two or three other black specks which we made out also to be boats. As daylight grew we made out more and more boats, three on one side of our ship and five on the other. Still later we picked up more.

"Here and there on the ocean's surface among the field of ice were bits of wreckage from the broken *Titanic,* and there were in sight many bergs eighty and ninety feet high. The passengers of the *Titanic* were taken aboard the *Carpathia* boatload by boatload up sea ladders.

MOST WOMEN HOISTED ABOARD

"The women, most of them, were hoisted to the decks of the *Carpathia* in swings but a few were hardy enough to climb aboard by the sea ladders. The ocean

90 WRECK OF THE TITANIC

all this time was calm as a lake and it was not a difficult task to take the excess passengers aboard.

"Some of the women helped out in the rowing in the lifeboats themselves."

Mrs. Ogden said that she saw the hands of Mrs. Astor, Mrs. John B. Thayer and Mrs. George D. Widener red from the oars. Most of the women were wet to the knees from the icy water that had slopped into the *Titanic's* lifeboats.

—*Indianapolis Star*
LEST WE FORGET

CHAPTER X

SORROW AND HONOR AND MEMORY EQUAL

HEROISM WAS UNIFORM AND UNIVERSAL AND NO DISTINCTIONS NEED BE DRAWN

There are differences between the statements of those rescued from the perished *Titanic*. There are contradictions as well as differences. The fact, however, but confirms the sincerity and the endeavor to be truthful of all who try to tell the story. Agreement on every detail would suggest collusion and impair faith in what was said.

Readers who bear these facts in mind will get at the substantial truth of the various accounts and draw the correct conclusions from them. The one and great conclusion to be drawn is that which proves the bravery and unselfishness of officers and crew and passengers, the fortitude of women, the consideration of all for the children, and the credit the entire story casts on the unselfishness of human beings in a sudden and concerted exchange of worlds.

If the tragedy is sorrow's crown of sorrow, the tragedy is likewise a justification of the claim of the lost to the honor as well as to the pity of the race and to the assurances they were as dear to the Heart of God as they will forever be to the chronicles and traditions of men. Every soul alone knows and can never fully tell its own grief. Every household alone realizes and can never fully tell its own loss. No riven heart can ever believe another's heart suffers woe like unto its woe. That is universal because natural. It is also in process of time consoling.

Equally true it is that there should be no comparisons instituted between exemplars of heroism where heroism was uniform and universal. Any one of us well knew friends who perished together, in one another's arms maybe. But others, too, know friends of theirs who met the same fate with the same courage. Comparison, contrast or competition of credit under such circumstances were revolting and impossible.

The men who have died for men have won the laurels of the race. The men who died for women are entitled to the love as well as to the laurels of the race. The men who died for little children are evermore shrined in the heart of Him "Who took the little children in His arms and blessed them," as He said, "For of such is the Kingdom of Heaven."

If there is any rose of distinction in the chaplet of memory, let it go to the husbands and wives who literally loved, lived and died together, each refusing to survive the other. For those dead the portals of Eternity swung wide open, but in the souls of those who went through them together must have been special joy, and for them well could be special honor and shall be

The equal and equally honored and equally mourned dead should have and will have equal remembrance among the living. For them let sudden death be held to have been the assured glory of those who did die or were ready to die that others might live.

"For this cause shall a man lay down even his life," said He who once laid down even His for His enemies. In this instance not a few surrendered their lives even for strangers. The Friend and Father of all the race has no rebuke for those made in His image who followed His example. God accepts them. Christ receives them. Humanity cannot forget them. The summons all must answer, and most of us alone, is answered with special pathos and power on the sea, in the night and in grouped comradeship, with the consciousness and comfort as time recedes and Heaven opens, that if for them who live for others earth is well, for them who die for others Eternity has an abundant entrance into love ineffable.

THE LAST WORD FROM THE TITANIC

"We rowed frantically away from the *Titanic* and were tied to four other boats. I arose and saw the ship sinking. The band was playing 'Nearer, My God to Thee.'"—*Mrs. W. J. Douton, a survivor, whose husband was drowned.*

>Nearer, my God, to Thee,
>   Nearer to Thee!
>E'en though it be a cross
>   That raiseth me;
>Still all my song shall be
>Nearer, my God, to Thee,
>   Nearer to Thee!
>
>Though like the wanderer,
>   The sun gone down,
>Darkness be over me,
>   My rest a stone;
>Yet in my dreams I'd be
>Nearer, my God, to Thee,
>   Nearer to Thee!
>
>There let the way appear
>   Steps unto heaven;
>All that thou sendest me
>   In mercy given;

WRECK OF THE TITANIC

Angels to beckon me,
Nearer, my God, to Thee,
  Nearer to Thee!

Then with my waking thoughts,
  Bright with thy praise,
Out of my stony griefs
  Bethel I'll raise;
So by my woes to be
Nearer, my God, to Thee,
  Nearer to Thee!

Or if on joyful wing
  Cleaving the sky,
Sun, moon and stars forgot,
  Upward I fly;
Still all my song shall be
  Nearer, my God, to Thee,
  Nearer to Thee.

CHAPTER XI

THE RESPONSIBILITY FOR FATAL SPEED

THE CAPTAIN WAS UNDOUBTEDLY CARRYING OUT INSTRUCTIONS OF THE OWNERS

The investigation of a committee of the United States Senate brought out all the material facts bearing upon the disaster that sent the *Titanic* and 1,595 persons to the bottom of the Atlantic. Mr. Bruce Ismay, managing director of the White Star Line, the first witness, deposed under oath that at the time of the collision the ship was not going at full speed. That is a matter of deduction from his testimony. "The ship's full speed was 78 revolutions. We did not make more than 72." The *Titanic* could steam between 22 and 23 knots an hour, so it is evident that her speed was at the rate of about 21 knots, and therefore high in an ice drift where bergs could be seen by daylight and might be encountered suddenly after dark.

It was a clear, starlight night, the sea was calm, and except for the presence of loose floes and masses of ice with submerged bases there was no reason why the *Titanic* should not have been making good time.

But the exception was very important. Obviously the great ship was proceeding at a high rate of speed under orders of the captain, who just as obviously was trying to carry out the instructions of his employers. If the *Titanic* was not as fast a ship as the *Lusitania* or the *Mauretania* she was expected to make a good record on her maiden trip, which could not be done unless she held to a prescribed route. It was certainly in the power of Mr. Ismay to have the *Titanic's* course changed to the south when dangerous ice was reported ahead. The warning had come by wireless from the *Amerika* the day before the disaster. But to take at once a more southerly course would have involved a loss in time of several hours at least on the maiden voyage of the great *Titanic*.

After the tragic event it seems criminal that the course was not changed if the new ship was to be driven on at a speed of 21 knots. The alternative was to proceed slowly through the ice field, but at a rate to keep her under perfect control. A steamship of the size of the *Titanic* must maintain a speed proportionately greater than the speed at which a vessel of half her tonnage can be handled in an emergency. What, then, is the explanation of her forging through ice-strewn water almost at maximum velocity? Can there be any doubt that the risk was not understood? Swiftly to condemn is to lose sight of the fact that the experience of captains of transatlantic liners with fields of ice, particularly with bergs partly submerged, is negligible. To the commander of the *Titanic,* a veteran who had made the passage hundreds of times, the conditions that destroyed his ship presented no perils requiring him to slow down to headway speed or to safe manœuvring speed. It was sufficient for him that the night was clear, that the ice was loose. He believed, as he had declared before he took charge of the ship, that she was unsinkable. A faith fatal in its consequences, but he knew nothing of the power of a great mass of floating ice to tear out the side of a 45,000-ton ship and smash in her watertight compartments. It is clear enough that the loss of the *Titanic* and the sacrifice of two-thirds of her passengers and crew was due more to ignorance and misplaced confidence than to criminal carelessness.

After the event the world knows that a fearful risk was taken that ought to have been avoided. It is the old painful story of implicit faith in experience that proved valueless and in judgment that was fallible. A thousand and a half lives seem to have been wantonly sacrificed, but to place the responsibility without mitigation is not as simple as it seems in the shadow of the awful disaster. The verdict will be pronounced unflinchingly, but let the investigation be deliberate and the evidence complete.—*New York Sun.*

CHAPTER XII

OTHER CONTRIBUTING CAUSES OF THE DISASTER

IN ADDITION TO LACK OF LIFEBOATS, CREWS DID NOT KNOW HOW TO MANAGE THOSE THEY HAD—ALSO FIRE RAGED IN COAL BUNKERS FROM THE START—INEXPERIENCED CREW

There was some criticism among the survivors of the *Titanic* crew's inability to handle the lifeboats. "The crew of the *Titanic* was a new one, of course," declared Mrs. George N. Stone of Cincinnati, "and had never been through a lifeboat drill, or any training in the rudiments of launching, manning and equipping the boats. Scores of lives were thus ruthlessly wasted, a sacrifice to inefficiency. Had there been any sea running, instead of the glassy calm that prevailed, not a single passenger would have safely reached the surface of the water. The men did not know how to lower the boats; the boats were not provisioned; many of the sailors could not handle an oar with reasonable skill."

NO BOAT DRILLS HELD

Albert Major, steward of the *Titanic,* admitted that there had been no boat drills and that the lifeboats were poorly handled.

"One thing comes to my mind above all else as I live over again the sinking of the *Titanic*," he said. "We of the crew realized at the start of the trouble that we were unorganized, and, although every man did his best, we were hindered in getting the best results because we could not pull together.

"There had not been a single boat drill on the *Titanic*. The only time we were brought together was when we were mustered for roll call about 9 o'clock on the morning we sailed. From Wednesday noon until Sunday nearly five days passed, but there was no boat drill."

The White Star liner *Titanic* was on fire from the day she sailed from Southampton. Her officers and crew knew it, for they had fought the fire for days.

This story, told for the first time on the day of landing by the survivors of the crew who were sent back to England on board the Red Star liner Lapland, was only one of the many thrilling tales of the first—and last—voyage of the *Titanic*.

"The *Titanic* sailed from Southampton on Wednesday, April 10, at noon," said J. Dilley, fireman on the *Titanic*, who lives at 21 Milton road, Newington, London, North, and who sailed with 150 other members of the *Titanic's* crew on the *Lapland*.

"I was assigned to the *Titanic* from the *Oceanic*, where I had served as a fireman. From the day we sailed the *Titanic* was on fire, and my sole duty, together with eleven other men, had been to fight that fire. We had made no headway against it.

"Of course, sir," he went on, "the passengers knew nothing of the fire. Do you think, sir, we'd have let them know about it? No, sir.

"The fire started in bunker No. 6. There were hundreds of tons of coal stored there. The coal on top of the bunker was wet, as all the coal should have been, but down at the bottom of the bunker the coal had been permitted to get dry.

"The dry coal at the bottom of the pile took fire, sir, and smoldered for days. The wet coal on top kept the flames from coming through, but down in the bottom of the bunker, sir, the flames was a-raging.

"Two men from each watch of stokers were told off, sir, to fight that fire. The stokers, you know, sir, work four hours at a time, so twelve of us was fighting flames from the day we put out of Southampton until we hit the iceberg.

"No, sir, we didn't get that fire out, and among the stokers there was talk, sir, that we'd have to empty the big coal bunkers after we'd put our passengers off in New York and then call on the fireboats there to help us put out the fire.

"But we didn't need such help. It was right under bunker No. 6 that the iceberg tore the biggest hole in the

Titanic, and the flood of water that came through, sir, put out the fire that our tons and tons of water had not been able to get rid of.

"The stokers were beginning to get alarmed over it, but the officers told us to keep our mouths shut—they didn't want to alarm the passengers."

Another story told by members of the *Titanic's* crew, was of a fire which is said to have started in one of the coal bunkers of the vessel shortly after she left her dock at Southampton, and which was not extinguished until Saturday afternoon. The story, as told by a fireman, was as follows:

"It had been necessary to take the coal out of sections 2 and 3 on the starboard side, forward, and when the water came rushing in after the collision with the ice the bulkheads would not hold because they did not have the supporting weight of the coal. Somebody reported to Chief Engineer Bell that the forward bulkhead had given way and the engineer replied: 'My God, we are lost.'

"The engineers stayed by the pumps and went down with the ship. The firemen and stokers were sent on deck five minutes before the *Titanic* sank, when it was seen that they would inevitably be lost if they stayed longer at their work of trying to keep the fires in the boilers and the pumps at work. The lights burned to the last because the dynamos were run by oil engines."

CHAPTER XIII

MORE OF THE TRAGEDY

DEATH WAITED FOR EVERYONE, RICH AND POOR ALIKE, ON THE ILL-FATED SHIP

George D. Widener, the wealthy Philadelphian, and Arthur L. Ryerson of New York went to their deaths like men, is the statement by Mrs. Ryerson to her brother-in-law, E. S. Ryerson, after her rescue. She says that when the women were put into the lifeboats they saw Mr. Ryerson and Mr. Widener standing behind the rail of the *Titanic,* both waving their arms, throwing kisses and calling farewell to their wives and children. They believed there were boats enough for all. Mrs. Ryerson had her two daughters, Susan and Emily B., and a young son, John B., in the boat with her.

AIR-TIGHT CHAMBERS PROVED DEATH CELLS

That fifty or more steerage passengers of the *Titanic* were immured in a steel prison from which escape was impossible with the closing of the air-tight compartment doors in the steerage deck forward of midships was the statement made by a member of the ship's crew and who himself verified the fact that escape had been shut off for these unfortunates.

WRECK OF THE TITANIC

To have opened the doors which shut off these steerage passengers from the decks and possible escape would have been to shorten the life of the ship, he declared, and hurry disaster on all of the hundreds crowded about the boat davits high above.

NO CHANCE FOR LIVES

"I know that fifty or more steerage passengers, whose quarters were on the same deck with the gloryhole used by the stewards of the second cabin, never got a chance for their lives," the informant said. "I know, because I nearly got caught myself by the closing of the water-tight doors leading from the working alley, which opened from the forward deck through to all the forepart of the ship.

"At the first shock all of the stewards in my gloryhole, forty all told, tumbled from their bunks and went out through the working alley to see what the trouble was: I heard some one give an order, 'Look out for the water-tight doors.' A minute later I started to go back to the glory-hole to get a life belt, the order having been passed out to all members of the crew to equip themselves with these belts.

STEEL DOORS SLAMMED

"I could not get back through the alley to the gloryhole because the water-tight doors had slammed tight across the passageway. There was no way around it.

WRECK OF THE TITANIC 105

There was no way for those on the other side of it, in the forepeak of the ship, to get out to open air.

"I know that none of the people from the steerage sleeping quarters beyond that water-tight door got out before it was shut, because they would have had to pass me in the alley, and none of them did. I spoke to one of the petty officers about the door being shut and all those people in there, and he said: 'Well, what can we do about it now? If those forward compartments hold, then the air in them will keep us up all the longer.'"

BELLBOYS AS WELL AS MILLIONAIRES

Among the many hundred of heroic souls who went bravely and quietly to their end were fifty happy-go-lucky youngsters shipped as bellboys or messengers to serve the first cabin passengers. James Humphries, a quartermaster, who commanded lifeboat No. 11, told a little story that shows how these fifty lads met death.

Humphries said the boys were called to their regular posts in the main cabin entry and taken in charge by their captain, a steward. They were ordered to remain in the cabin and not get in the way. Throughout the first hour of confusion and terror these lads sat quietly their benches in various parts of the first cabin.

Then, just toward the end, when the order was passed around that the ship was going down and every man was free to save himself if he kept away from the lifeboats in which the women were taken, the bellboys scattered to all parts of the ship.

Humphries said he saw numbers of them smoking cigarettes and joking with the passengers. They seemed to think that their violation of the rule against smoking while on duty was a sufficient breach of discipline.

Not one of them attempted to enter a lifeboat.

Not one of them was saved.

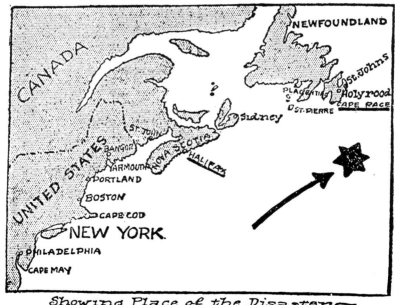

Showing Place of the Disaster

CHAPTER XIV

ODDITIES OF THE WRECK

FATE PLAYED SOME STRANGE FREAKS ALONG WITH THE HORROR—MONEY LESS VALUABLE THAN ORANGES

One of the cabin passengers of the *Titanic,* Maj. A. G. Peuchen of Toronto, left more than $300,000 in money, jewelry and securities in a box in his cabin when he left the ship. He went back to his cabin for the box, but decided to take instead three oranges.

"The money seemed to be a mere mockery at that time," said the major. "The only trinket I saved was a little pin which I remembered had always brought me luck. I picked up the pin and three oranges instead of the money and the documents."

Maj. Peuchen, who is president of the Standard Chemical Company of Canada and vice commodore of the Royal Canadian Yacht Club, was thrust into one of the boats by the captain and ordered to man an oar.

DEMANDED A BATH

G. Wikeman, the *Titanic's* barber, was treated for bruises. He declared that he was blown into the water by the second explosion on the *Titanic,* after her collision with the iceberg.

A passenger who was picked up in a drowning condition caused grim amusement on the *Carpathia* by demanding a bath as soon as the doctors were through with him.

JUMPED FROM THE DECK

Storekeeper Prentice, the last man off the *Titanic* to reach the *Carpathia,* swam about in the icy water for hours, but soon was restored. He said he had leaped from the *Titanic's* poop deck.

Mrs. James Baxter and her daughter, Mrs. P. C. Douglas of Montreal, Canada, when rescued were wearing the evening dresses that they had on at the Sunday night concert on the *Titanic,* having lost all their other wearing apparel.

A ROMANCE OF THE WRECK

In the midst of death and horror, Cupid played a little game and won.

One of the girl survivors of the *Titanic,* Miss Marion Wright of Somerset, England, was married in New York the day after landing, to Arthur Woolcott of Cottage Grove, Ore. She came alone from her home in England to meet her fiance and he had been in New

York for nearly a week anxiously awaiting her arrival. The pair were schoolmates in England and became engaged before Mr. Woolcott left to become an Oregon fruit grower.

GETTIN' THE LESSON

—*Indianapolis News*

Copyright Underwood & Underwood
A GROUP OF RESCUED PASSENGERS BEING SUPPLIED WITH WRAPS ABOARD CARPATHIA

CHAPTER XV

THE TERROR OF THE SEAS

By Fred S. Miller.

There is one, and but one, danger to navigation against which the ingenuity of navigators is absolutely powerless, and this danger is formed by the vast icebergs—floating ice-prairies, some of them—which every month in the year, but more particularly in the winter months, are sent in shoals from the Arctic and the Antarctic regions to float down the currents of the ocean until they are finally melted and mingled with warm waters. A brief account of the origin of these marine monsters, their action and the manifold dangers they present to sailors, will be of interest.

Greenland is the breeder of the iceberg for the northern seas. Greenland is a mysterious continent on which no vegetable life can endure. Its exact limits have never yet been traced, but is known to be comparatively flat, though covered to immense depths by snow and ice. This snow and ice forms constantly throughout the year, and has so formed since prehistoric times. It heaps up so that the surface of Greenland may be roughly compared to a vast hill. The enormous weight

LAUNCHING THE S.S. TITANIC

of this constantly forming ice causes movements of the masses from the center to the sea, and thus the glaciers are formed—vast processions of granite-hard ice which "flow" very slowly but irresistibly and for vast extents down to the water.

The size of these great moving plains is indeed almost unbelievable. The Humboldt glacier is sixty miles broad, its walls rise three hundred feet from the place where it meets the sea, and as to its depth inland it has been plumbed for half a mile. Every year it sends out over the ocean a mass whose area is greater than that of the State of New Jersey.

Another of the great Greenland glaciers, called the Jacobshaven glacier, is two thousand feet broad and one thousand feet high, and its output to the sea is estimated as being over 400,000,000,000 cubic feet of ice yearly.

Thousands of miles of this matter are constantly being emptied into the ocean, the rate of progress being about forty-two feet a day. Immense masses of solid ice creep along the shore, at the water's edge presenting a vertical face of steel-blue ice hard as flint, against which dash the angry waves of the Arctic. Out this ice pushes, day after day, until finally its own weight or the action of the water causes vast sections to break off with a roar like that of a thousand thunder claps and with a disturbance in the ocean that could only be compared to the commotion caused by the birth of a new island. Thus born, the berg floats gently down the currents for the Grand Banks of Labrador, where the fogs and mists that continually wreathe that region, shut the icy menace from view of the anxious mariner frequently until it is too late for him to turn his vessel to avoid them. In such weather it is of no help for the lookout on the tops that the iceberg frequently towers hundreds of feet into the air. It cannot be seen for the dense blanket of fog that shuts out sight and shuts out sound, so that even the wash of the waves dashing against the base of the approaching destroyer cannot be heard. Only by the cold radiated from it may its presence be guessed, but if the wind is blowing from the vessel to the berg the temperature cannot be felt lowering until the boat is so near that it is impossible to turn it before the crash comes.

Again, many of these great masses cannot be seen above the surface of the sea as they only extend comparatively a few feet into the air. Nevertheless eight-ninths of the berg is always under water, so that, especially at night, a vast plateau of ice may be gliding towards a steamer and giving no indication of its presence.

The steamer *Saale,* coming over the same course as that taken by the *Titanic,* was in 1890 subjected to almost the same experience although she escaped as by a miracle. Rushing along in the midnight gloom its

WRECK OF THE TITANIC 115

path was suddenly barred by a black rampart of steely ice, 100 feet high. The lookout gave timely warning, the engines were reversed and the helm put hard aport, so that the steamer barely crunched along over the submerged foot of the berg, bumping heavily a few times and being shot off into deep water sidewise so that the coal and cargo were shifted. This listed the vessel heavily, in which plight she proceeded slowly to port, her starboard rail barely clearing the water.

The *Normania,* in 1900, had a similar experience. It turned just in time to avoid a direct impact with an immense berg, but it ran alongside of the floating mountain, shearing its sides and showering itself with ice scraped off, which loaded the decks.

But these are merely lucky escapes. By far the greater number of vessels, once they touch the frightful mass of beetling crag and jagged base, are lost on the moment of impact, the passengers being lucky if they have the time and the boats to escape with. The record of the sea is heavy with the account of gallant ships that perished, some with all on board. Many is the number that went down and were never heard from nor a vestige of them seen, but which were supposed to have been overborne by icebergs. Until very recent years the wireless telegraph was unheard of and ships suddenly overtaken could not communicate their plight but must vanish without leaving a record that they had ever been. In this way went the *Ismalia*, the *Columbo*, the *Homer, Zanzibar, Surbiton* and *Bernicia*, and to this day no light has been thrown on the mystery of their loss. Of course there are many more similar cases, any year of the last twenty being prolific with instances of these mysteriously disappearing ships.

The only vessel that can hope to escape destruction by contact with an iceberg is the especially strengthened ship built for Arctic exploration. Ships like the *Fram* of Amundsen, or Peary's ship, are proof against even a head-on collision as they are very strong and very light. But an ocean liner is especially vulnerable. Going at the rapid speed that is nearly always maintained on these palatial ships, and with their enormous weight and displacement and their comparatively weak structure, the momentum which they acquire shatters them like glass when it is brought to an instant stop against a sluggish-moving mass say a mile long, two hundred feet above the water, 1,600 feet below the water, of a weight incalculably great and of a hardness like granite.

Much time has been spent and many efforts have been made to devise some instrument or discover some means whereby the presence of an approaching iceberg might be detected, but so far little progress has been made toward perfecting anything that at all answers the requirements. The towering berg can of course be

WRECK OF THE TITANIC

seen for miles unless hidden by the fog, but what of the immense masses that lie scarcely visible in the water yet wholly destructive of whatever ship shall hurl itself upon that jagged floating reef of ice-coral? Ships that run on top of such bergs break literally in two, as their keels are not made to sustain a strain of balancing or "teetering" as the ship does when it runs upon the uneven surface of the berg.

But if icebergs are terrible they are beyond doubt among the most beautiful and superb manifestations of nature. Think of a mass of glittering minarets and towers, of domes, arches, collonades, spires and special forms and features of its own uniquely beautiful—think of such a mass irradiant with a thousand variations of the rainbow hues and flashing in the sunlight of a northern summer day; think of a landscapeful of this delirious beauty, a bulk as large as the State of Rhode Island, moving majestically to the open ocean, breaking into mysterious peals of thunder as it dominates the sea! Perhaps it will receive and override some goodly vessel in its unruffled progress from the cold inconceivable which brought it forth. Perhaps the luckless voyagers will view its dreadful shape with an awe that will impel them rather to perish in the deep than to endevor to seek refuge on the sheer and frigid walls that have o'erborne their ship. But presently the enormous edifice of ice itself shall sink and perish in the sea, merged with the enervating waters of the Gulf Stream—o'erborne as all things are and set to uses new by that emanation called by the learned "the opposition of forces" and by the wise called God, which keeps His ministering universe in equipoise and holds its balance true.

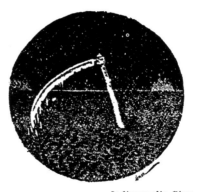

—*Indianapolis Star*
THE TOLL OF THE SEA